STEFFEN WOLF

OPTIMIZATION PROBLEMS IN SELF-ORGANIZING NETWORKS

Bibliografische Information der Deutschen Nationalbibliothek

Die Deutsche Nationalbibliothek verzeichnet diese Publikation in der
Deutschen Nationalbibliografie; detaillierte bibliografische Daten sind
im Internet über http://dnb.d-nb.de abrufbar.

ISBN 978-3-8325-2661-0

Logos Verlag Berlin GmbH
Comeniushof, Gubener Str. 47,
10243 Berlin
Tel.: +49 (0)30 42 85 10 90
Fax: +49 (0)30 42 85 10 92
INTERNET: http://www.logos-verlag.de

OPTIMIZATION PROBLEMS IN SELF-ORGANIZING NETWORKS

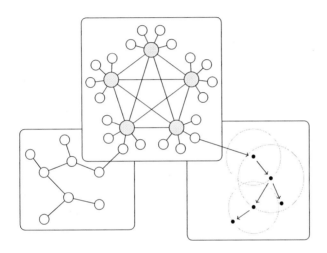

Vom Fachbereich Informatik der Universität Kaiserslautern
zur Verleihung des akademischen Grades
Doktor der Naturwissenschaften (Dr. rer. nat.)
genehmigte Dissertation von

STEFFEN WOLF

Wissenschaftliche Aussprache
28. Juni 2010
Dekan des Fachbereichs
Prof. Dr. rer. nat. Karsten Berns
Vorsitzender der Promotionskommission
Prof. Dr. rer. nat. Hans Hagen, TU Kaiserslautern
Berichterstatter
Prof. Dr.-Ing. Peter Merz, FH Hannover
Prof. Dr.-Ing. Jens B. Schmitt, TU Kaiserslautern
Zeichen der TU Kaiserslautern im Bibliotheksverkehr: D 386

ABSTRACT

Networks are part of our daily life. Whether they are road networks, networks of gas pipelines or computer networks, they often pose challenging optimization problems. The construction of such networks usually induces costs which should be minimized. There are also costs for using established networks, such as fuel costs, travelling time, or message delay in computer networks. When considering wireless networks, the transmission of a message also creates costs in form of energy. As the available energy is often limited, applying algorithms to reduce the energy consumption can increase the wireless network's lifetime.

This thesis presents algorithms to solve such optimization problems. Especially considered are networks of autonomous nodes, e. g. computer networks or wireless networks, as this allows to design algorithms that enable the participants in these networks to collectively solve the optimization problems, thus laying grounds for distributed optimization of networks. Using these algorithms, the network can be modelled as a self-optimizing network.

Starting from the mathematical formulations for the optimization problems, various global view optimization algorithms are presented. These algorithms are used to evaluate the possible gain of optimization. Distributed algorithms are presented to create a self-optimizing network. The performance of these distributed algorithms is evaluated using the results of the corresponding global view algorithm.

Three optimization problems are considered in this thesis: The Minimum Routing Cost Spanning Tree (MRCST) and the Super-Peer Selection Problem (SPSP) are examples from computer communication networks, dealing with the reduction of message delay in Peer-to-Peer (P2P) systems. This not only brings an advantage for the respective P2P application, but also helps reducing the load on the underlying network. The topologies created by the proposed distributed algorithms can be used e. g. to create a P2P desktop grid. The third optimization problem is related to wireless networks. The Range Assignment Problems (RAPs) are motivated by the need to reduce the energy consumption for wireless communication. There are a number of variants of RAPs, each assuming another communication scheme. Two of these variants are used in this thesis.

ZUSAMMENFASSUNG

Netzwerke sind Teil unseres täglichen Lebens. Sie treten auf in Form von Straßennetzen, Ferngasnetzen oder Computernetzen und bringen oft interessante Optimierungsprobleme mit sich. Die Errichtung eines solchen Netzes erfordert Kosten, die minimiert werden sollten. Auch später entstehen Kosten, etwa Transportkosten, Fahrzeiten, oder Nachrichtenlaufzeiten in Computernetzen. In drahtlosen Netzen erzeugt bereits die Nachrichtenübermittlung Kosten in Form von Energie. Da die verfügbare Energie oft begrenzt ist, können hier Algorithmen zur Reduzierung des Energieverbrauchs die Lebensdauer des Netzes erhöhen.

Diese Arbeit stellt Algorithmen für derartige Optimierungsprobleme vor. Von besonderem Interesse sind Netzwerke autonomer Knoten, etwa Computernetze und drahtlose Netze, da diese die Entwicklung von Algorithmen erlauben, mit denen die Teilnehmer des Netzes diese Optimierungsprobleme selbst lösen können. Damit werden Grundlagen für die verteilte Optimierung in Netzwerken gelegt. Das Netzwerk kann somit auch als selbst-optimierendes Netzwerk angesehen werden.

Ausgehend von den mathematischen Grundlagen werden Optimierungsalgorithmen vorgestellt, die globale Sicht benötigen. Diese helfen, den möglichen Nutzen einer Optimierung aufzuzeigen. Verteilte Algorithmen werden vorgestellt, die selbst-optimierende Netze erzeugen. Die Ergebnisse der verteilten Algorithmen werden auf der Basis der Ergebnisse der globalen Optimierungsalgorithmen ausgewertet.

Drei Optimierungsprobleme werden in dieser Arbeit behandelt: Der *Minimum Routing Cost Spanning Tree* (MRCST) und das *Super-Peer Selection Problem* (SPSP) sind Beispiele aus dem Bereich der Computernetze. Hier sollen die Nachrichtenlaufzeiten in Peer-to-Peer-Netzen reduziert werden, da dies nicht nur performantere Kommunikation im P2P-System erlaubt, sondern auch das zugrundeliegende Netzwerk schont. Die vorgestellten Algorithmen können etwa genutzt werden, um ein P2P-Desktop-Grid aufzubauen. Das dritte Problem stammt aus dem Bereich der drahtlosen Netze. Die *Range Assignment Problems* (RAPs) befassen sich mit der Verringerung des Energieverbrauchs bei drahtloser Nachrichtenübertragung. Von den verschiedenen Varianten der RAPs, die jeweils verschiedene Annahmen über die Kommunikation im Netzwerk treffen, werden zwei in dieser Arbeit betrachtet.

ACKNOWLEDGEMENTS

This thesis would not have been possible without the support of many friends and colleagues. I would like to take this opportunity to thank them individually.

A big thank-you goes to my advisor and mentor Prof. Peter Merz, who encouraged me to follow the paths of research and nudged me back to the original topic whenever this was necessary. His vast knowledge on memetic algorithms was a huge support, as well as his contacts to the optimization community. Through his support I had the pleasure to meet this community at various interesting conferences, such as the EvoCOP.

Many thanks also to my second supervisor and reviewer Prof. Jens B. Schmitt. He made sure that my somewhat theoretical research still had connections to practical problems. Thanks also for presenting and supporting my PhD thesis at the faculty council, and for taking me in after the Distributed Algorithms Group scattered.

Thomas Fischer was a big help with his knowledge about heuristics, Linux and LaTeX. His thesis inspired me a lot. He also helped me navigate through the PhD programme and brought me into contact with the local geeks.

Matthias Priebe suffered a lot from my attempts to explain complex mathematical ideas. In exchange, he amazed me with his huge active English vocabulary. I think, we made a good team. The success of our work is proof of that.

Michael Weiser, Thomas Steinbach, Eugen Weber and Marcus Kunze made sure, I did not overwork myself, by taking me on hiking, cycling or canoeing vacations.

Many thanks as well to those who discussed research related topics with me: Kerstin Bauer should be mentioned, as well as Tanja Meyer and Johannes Kloos. Thanks also to the countless proofreaders, your suggestions helped to improve this thesis a lot.

Finally, I want to thank all the Wolfs for their support during my studies and beyond, for their confidence in me, and for their patience when I was preparing this thesis for publication.

PUBLICATIONS

Some ideas and figures have appeared previously in the following publications:

[112] Peter Merz and Steffen Wolf. Evolutionary Local Search for Designing Peer-to-Peer Overlay Topologies based on Minimum Routing Cost Spanning Trees. In Thomas Philip Runarsson, Hans-Georg Beyer, Edmund Burke, Juan J. Merelo-Guervós, L. Darrell Whitley, and Xin Yao, editors, *PPSN 2006*, volume 4193 of *LNCS*, pages 272–281. Springer, Heidelberg, 2006. (Cited on pages 47 and 60.)

[113] Peter Merz and Steffen Wolf. TreeOpt: Self-Organizing, Evolving P2P Overlay Topologies Based On Spanning Trees. In Torsten Braun, Georg Carle, and Burkhard Stiller, editors, *Kommunikation in Verteilten Systemen (KiVS 2007) – Industriebeiträge, Kurzbeiträge und Workshops*, pages 231–242. VDE Verlag, Berlin, 2007. (Cited on page 82.)

[184] Steffen Wolf and Peter Merz. Efficient Cycle Search for the Minimum Routing Cost Spanning Tree Problem. In Peter Cowling and Peter Merz, editors, *EvoCOP 2010 – Tenth European Conference on Evolutionary Computation in Combinatorial Optimization*, volume 6022 of *LNCS*, pages 276–287. Springer, Heidelberg, 2010. (Cited on page 60.)

[180] Steffen Wolf. On the Complexity of the Uncapacitated Single Allocation p-Hub Median Problem with Equal Weights. Internal Report 363/07, University of Kaiserslautern, Kaiserslautern, Germany, 2007. (Cited on page 223.)

[181] Steffen Wolf and Peter Merz. Evolutionary Local Search for the Super-Peer Selection Problem and the p-Hub Median Problem. In Thomas Bartz-Beielstein, María José Blesa Aguilera, Christian Blum, Boris Naujoks, Andrea Roli, Günther Rudolph, and Michael Sampels, editors, *HM 2007 – 4th International Workshop on Hybrid Metaheuristics*, volume 4771 of *LNCS*, pages 1–15. Springer, Heidelberg, 2007. (Cited on pages 92 and 107.)

[115] Peter Merz, Matthias Priebe, and Steffen Wolf. A Simulation Framework for Distributed Super-Peer Topology Construction Using Network Coordinates. In Didier El Baz, Julien Bourgeois, and François Spies, editors, *16th Euromicro Conference on Parallel, Distributed and Network-based Processing*, pages 491–498. IEEE Computer Society, Los Alamitos, 2008. (Cited on page 135.)

[116] Peter Merz, Matthias Priebe, and Steffen Wolf. Super-Peer Selection in Peer-to-Peer Networks using Network Coordinates. In *Proceedings of the 3rd International Conference on Internet and Web Applications and Services (ICIW 2008)*, pages 385–390. IEEE Computer Society, Los Alamitos, 2008. (Cited on page 133.)

[118] Peter Merz, Steffen Wolf, Dennis Schwerdel, and Matthias Priebe. A Self-Organizing Super-Peer Overlay with a Chord Core for Desktop Grids. In Karin Anna Hummel and James P.G. Sterbenz, editors, *Proceedings of the 3rd International Workshop on Self-Organizing Systems (IWSOS)*, volume 5343 of *LNCS*, pages 23–34. Springer, Heidelberg, 2008. (Cited on page 152.)

[160] Sameh Al-Shihabi, Peter Merz, and Steffen Wolf. Nested Partitioning for the Minimum Energy Broadcast Problem. In Vittorio Maniezzo, Roberto Battiti, and Jean-Paul Watson, editors, *LION 2007 II. Selected Papers*, volume 5313 of *LNCS*, pages 1–11. Springer, Heidelberg, 2008. (Cited on pages 186 and 187.)

[182] Steffen Wolf and Peter Merz. Evolutionary Local Search for the Minimum Energy Broadcast Problem. In Jano van Hemert and Carlos Cotta, editors, *EvoCOP 2008*, volume 4972 of *LNCS*, pages 61–72. Springer, Heidelberg, 2008. (Cited on page 186.)

[183] Steffen Wolf and Peter Merz. Iterated Local Search for Minimum Power Symmetric Connectivity in Wireless Networks. In Carlos Cotta and Peter Cowling, editors, *EvoCOP 2009 – Ninth European Conference on Evolutionary Computation in Combinatorial Optimization*, volume 5482 of *LNCS*, pages 192–203. Springer, Heidelberg, 2009. (Cited on page 204.)

[185] Steffen Wolf, Tom Ansay, and Peter Merz. A Distributed Range Assignment Protocol. In Thrasyvoulos Spyropoulos and Karin Anna Hummel, editors, *IWSOS 2009*, volume 5918 of *LNCS*, pages 226–231. IFIP International Federation for Information Processing, 2009. (Cited on page 213.)

CONTENTS

LIST OF FIGURES

LIST OF TABLES

ACRONYMS

AMLS	Ahuja-Murty Local Search
APSP	All-Pairs-Shortest-Path
ACO	Ant Colony Optimization
B&B	Branch and Bound
BIP	Broadcast Incremental Power
DLB	Don't Look Bit
DLM	Don't Look Marker
EA	Evolutionary Algorithm
EFS	Edge and Fork Switching Search
ELS	Evolutionary Local Search
ES	Edge Switching Search
ES2	Double Edge Switching Search
EWMA	Embedded Wireless Multicast Advantage
GA	Genetic Algorithm
GNP	Global Network Positioning
FDC	Fitness Distance Correlation
ILS	Iterated Local Search
IPP	Incremental Power: Prim
IPK	Incremental Power: Kruskal
LESS	Largest Expanding Sweep Search
LP	Linear Programming

LS	Local Search
MA	Memetic Algorithm
MEB	Minimum Energy Broadcast
MIP	Mixed Integer Programming
MMAS	Max-Min Ant System
MRCST	Minimum Routing Cost Spanning Tree
MPSCP	Minimum Power Symmetric Connectivity Problem
MST	Minimum Spanning Tree
NP	Nested Partitioning
OCST	Optimum Communication Spanning Tree
P2P	Peer-to-Peer
PIS	Proximity Identifier Selection
PRS	Proximity Route Selection
PTAS	Polynomial Time Approximation Scheme
RAP	Range Assignment Problem
RTT	Round Trip Time
SA	Simulated Annealing
SAT	Boolean Satisfiability Problem
SPSA	Super-Peer Selection Algorithm
SPSP	Super-Peer Selection Problem
SPT	Shortest Path Tree
SRAP	Symmetric Range Assignment Problem
ST	Subtree Moving Search
TS	Tabu Search

1

INTRODUCTION

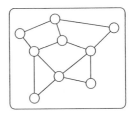

The rather abstract mathematical concept of a graph can be found in many aspects of real life. Graphs exists in form of communication networks, transport infrastructure, such as roads or air travel networks, networks of gas pipelines, electrical circuits or even supply chains.

graphs

When these networks are used or when they are constructed, costs have to be paid. These can be costs for building pipelines, paving roads or installing cables. There are also indirect costs. For an air traffic network, the cost of the link can be derived from the fuel needed for travelling the link. An even more indirect concept of cost is the travelling time. When calculating a route through a national road network, this is often the most important cost that has to be considered. In Peer-to-Peer (P2P) overlay networks, the message delay is usually the only cost, since the physical network already exists and the end user does not have to pay for the bandwidth. In all these examples, it is reasonable to assume that costs should be minimized.

costs

As soon as the costs can be specified, the construction of the network can be modelled as an optimization problem. A solution for this optimization problem gives a construction scheme for the network, or a list of directions, which when followed exactly minimize the cost for setting up the network or travelling to the desired location. Reasonable solutions can also be found when the costs for each individual link, such as a road or a flight connection, cannot be specified exactly. A good estimation for the cost is often enough. In many routing algorithms for road networks, the usual speed values for a road are defined by categorizing the roads. Here, less information leads to worse solutions. In some cases, global knowledge over the network is not achievable. This

optimization

happens, e. g. if a participant in the network tries to optimize or improve the network. This participant only has local knowledge. It could be a computer in a P2P system, a node in a wireless network, or an air traffic controller that only sees the traffic in his control area. The road example does not apply anymore, unless we assume a routing algorithm that needs to include unknown information about road blocks or traffic jams and model each car as a part of the network.

distributed optimization

In this thesis, we lay grounds for distributed optimization of networks. We will discuss mathematical foundations, global view optimization, as well as distributed approaches. Out of the vast field of optimization problems in networks, we picked three examples: The Minimum Routing Cost Spanning Tree (MRCST) and the Super-Peer Selection Problem (SPSP) are examples from computer communication networks. The set of Range Assignment Problems (RAPs) is an example in wireless communication networks. The main application of these examples lies in computer networks. However, applications can also be found in other areas: e. g. air travel, mail delivery, or all kinds of transport.

The motivation behind the MRCST and the SPSP are P2P topologies. P2P systems can be used for various purposes. The first application that made P2P systems famous was file sharing, however, P2P systems can also be used for content distribution, multimedia streaming, telephony, or resource sharing. A very interesting application can be found in desktop grids. These desktop grids were designed to make use of the idle computational resources of modern desktop computers. They allow to distribute computational tasks to a vast number of workers. These desktop grids benefit from the scalability of P2P systems. They also benefit from short message delays, as the initiator of a job wants to quickly find suitable workers, and the workers of a cooperative job need to exchange messages. Although the focus of this thesis lies on the optimization of the network topology, the approaches presented in this thesis can be used to create a desktop grid.

desktop grids

wireless networks

The motivation behind the RAPs is to reduce the energy consumption in wireless networks. These networks can also be used for various purposes. There are examples of large scale sensor networks detecting fire, or monitoring movements in a building. The so-called sensor and actor networks also react on perceived information. Wireless networks can also be used for communication, e. g. for establishing communication links to a rescuer team in a disaster area where the main communication lines have broken down. Again, this thesis does not provide application specific algorithms, but focusses on topology optimization, especially on energy consumption reduction.

This thesis is structured in five chapters. This first chapter gave a brief introduction. In the next chapter, we give an overview of optimization, heuristics and self-organizing networks, and provide definitions and explanations for the terms and concepts that are used in this thesis. The following three main chapters present our work on each of the three optimization problems. In the conclusion chapter, we summarize successful techniques for optimizing self-organizing networks and show connections between the approaches for these three problems.

structure

_2_

BASIC CONCEPTS

Since this thesis builds a bridge between global optimization and distributed algorithms, we will make use of specific terms from both research areas. This section gives an overview of the terms and techniques used. The following four sections describe networks, optimization, heuristics, and distributed algorithms.

2.1 NETWORKS

Networks exist in different forms. The introduction already mentioned transport infrastructure or telephone networks, however, in this thesis we mainly concentrate on computer networks. We will examine two variants: wireless networks and computer networks that are connected via cables or optical fibres.

2.1.1 Graph Representation

A network can usually be modelled as a *graph* $G = (V, E)$. The set V *graph*
gives the *nodes* in the network, and the set E gives the *edges* or links in the network. In the problems considered in this thesis, a node is either a computer in a computer network or a wireless node in a wireless network. Both have means of communication to other nodes, which means, they can establish links. We usually assume that every node can communicate with every other node, such that $E = V \times V$. However, there may be restrictions in the network that prevent the creation of some connections, such that $E \subseteq V \times V$.

In addition to the sets of nodes and edges, the graphs in this thesis also provide a non-negative *distance function* $d : E \to \mathbb{R}^+$. This func- *distance*
tion gives the distance between any pair of connected nodes or the cost *function*
for travelling this distance. Using this distance function, a graph that has missing edges can be transformed into a fully meshed graph by setting the distance for these edges to $d = \infty$.

A graph may be *directed* or *undirected*. In the context of P2P net- *directed vs.*
works, we usually assume an undirected graph, as P2P networks of- *undirected*

ten rely on bi-directional communication. An undirected graph can be modelled by using two edges (i, j) and (j, i) for the link between a node i and j, one for each direction, or by using the pair $\{i, j\}$ to denote the connection.

subgraph

These graphs define the possible edges of a network. However, a P2P system or a wireless network usually only uses a subset of these edges. This defines a *subgraph* $G' = (V', E')$, with $V' \subseteq V$ and $E' \subseteq E$. In the problems considered in this thesis, the subgraphs cover all nodes of the original graph. We use the term *topology* to refer to the subgraph used by the P2P system or the wireless network, i. e. all established links.

routes

The topologies are used for communication. If there is no direct link between two nodes, an indirect route can be used. A *route* is a sequence of connected links. Nodes along these routes are asked to forward the message to the destination. The cost for this route can be calculated as the sum of costs for each individual link. As all node pairs in the network may need to communicate, the topology is required to provide routes between all node pairs. If any two nodes in the network cannot communicate with each other, the network is said to be *partitioned*.

2.1.2   Topology Optimization

The problems considered in this thesis ask for a topology that minimizes a specific cost function. A simple example for this cost function is the sum of the costs for all used edges. This goal can be achieved by firstly reducing the number of edges in the graph. The minimal number of edges to connect a graph with n nodes is $n-1$, which builds a *tree*. The second step is to select those $n-1$ edges that use minimal cost for connecting all nodes. The result is a *Minimum Spanning Tree* (MST). It can be found easily using the algorithms of Prim [145] or Kruskal [94].

MST

SPT

Another important tree is a *Shortest Path Tree* (SPT). It consists of the shortest paths from one specified node to all other nodes. It can also be found easily using Dijkstra's algorithm [44].

tree
representation

Trees can be represented by their set of edges. However, a faster representation, especially when considering frequent tasks such as finding a route in a tree, can be achieved when using a *parent notation*. For each node i, this representation stores its parent $p(i)$. This is especially useful when the tree contains a special node $v^* \in V$, which is used as the root node in the tree and has no parent. However, any arbitrarily chosen node can also be used if no special node is present in the network. Using this representation, a route to this root node can be found

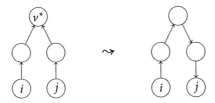

Figure 1: Tree representation using parent pointers. The arrows point to the parent of each node. These parent pointers can be used to find the path to the root node v^*. Every other route can be found by changing the directions of the arcs on the route from the desired destination to the root node, thus moving the root node to the desired destination.

by traversing the parent pointers. Routes to other nodes can be found by moving the root to these nodes, i. e. changing the direction of the links on the path from these nodes to the former root node, as Figure 1 shows.

The graph model abstracts from a large number of properties for real networks. Each link is represented only by a scalar cost. This could be the average message delay over this link or the energy needed to transmit a message. In reality, the link has many more properties, such as bandwidth, average packet loss, congestions, interference, or jitter. Also, the model does not include costs at the individual nodes. However, when these costs can be assumed equal for each node, they can be disregarded, as all nodes are used in the topology.

model limitations

2.1.3   Wireless Networks

Wireless networks can also be represented using a graph, however, for some applications a transmission range model is more appropriate [86, 152]. In wireless networks, the communication is not limited to a certain link, instead a local broadcast is used to transmit messages.

The energy needed for successfully transmitting a message to another node depends on the distance d to the receiver and on the medium. There are various models to describe the relation between distance and transmission power [152]. Common to all models is that they describe the *path loss factor PL*, which gives the quotient P_t/P_r of trans-

transmission energy

mit power P_t used at the sender and the signal strength P_r received at the receiver. The *free space propagation model* defines this path loss as

*free space
propagation
model*

$$PL(d) = \frac{(4\pi)^2 \cdot L}{G_t \cdot G_r \cdot \lambda^2} \cdot d^2 \qquad (2.1)$$

where G_t and G_r are the transmitter and receiver antenna gain, L is the system loss factor unrelated to propagation, λ is the wavelength of the radio signal, and d is the distance between transmitter and receiver.

The *two-ray ground model* gives another path loss factor:

*two-ray ground
model*

$$PL(d) = \frac{1}{G_t \cdot G_r \cdot h_t^2 \cdot h_r^2} \cdot d^4 \qquad (2.2)$$

Here, h_t and h_r give the transmitter and receiver antenna height over ground. In this model, the reflections of the radio transmission on the Earth's surface are assumed to improve the reception of the signal. Other models are also possible, where the reflection around the Earth's ionosphere is considered.

The *log-distance path model* combines the previous models in one simple approximative formulation:

$$PL(d) \propto d^\alpha \qquad (2.3)$$

The relation between transmission energy p and the distance can then be described using the following equation:

$$p = d^\alpha \qquad (2.4)$$

The parameter α is called *distance-power gradient*, or *path-loss exponent*. In free-space environments, it has a value of 2. Shadowed areas, obstacles, or in-building scenarios increase this value. It can take values of up to 6, depending on the environment. It can also take lower values in the case of constructive interference. Some common values are shown in Table 1.

Equation (2.4) can also be used to determine the set of receiving nodes for a transmission. All nodes that are within the transmission range can receive this message. The message may be attributed with a destination ID, allowing messages to be addressed to specific nodes. The graph model still applies, just the set of established links is determined by the transmission ranges of the individual nodes. If each node uses an individually chosen transmission range, the established links

*transmission
range*

Environment	α
Free space	2
Urban area	2.7–3.5
Indoor (line of sight)	1.6–1.8
Indoor (no line of sight)	4–6

Table 1: Distance-power gradient values for different environments. The values are empirical values taken from [152].

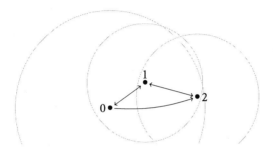

Figure 2: Wireless network example. The transmission ranges of the wireless nodes are drawn as grey circles. The possible links are indicated as arcs. In this example, node 2 does not have a sufficient transmission range to reach node 0.

may be uni-directional. A node with a higher transmission range may reach another node that has a lower transmission range. An example is illustrated in Figure 2, where node 2 can receive transmissions from node 0, but cannot reach it. Such links may be used for communication, however, they also cause problems, as the receiver cannot use this link to acknowledge the successful reception of a message. In the example, node 2 could ask node 1 to forward an acknowledgement packet.

Again, this model abstracts from some properties of wireless networks. A real wireless node also uses energy when encoding or decoding a transmission, when performing other computations, or when using other tools, such as a sensor. There may also be interferences, which cause retransmissions, and thus increase the necessary transmission energy for a link.

model limitations

2.2 OPTIMIZATION

Optimization deals with reducing costs or maximizing a gain. The problem of finding an MST can be formulated as an optimization problem. Optimization is not limited to graph problems, but since this thesis focusses on optimization problems in networks, there will be only few examples of other optimization problems.

2.2.1 *Definitions*

an optimization problem

Michael R. Garey and David S. Johnson define an *optimization problem* in [57] as a problem that asks for a structure with minimal cost among all such structures. In a way, the optimization problem asks for minimizing or maximizing a function, while observing a set of constraints.

an example

In the MST, this cost function is given by the sum of the costs for all used edges. The constraints of the MST are to use only edges of the given network and to select a set of edges that connects all nodes in the network. This defines the structure of the MST as a spanning tree.

formulation

Optimization problems are usually formulated as mathematical problems. The function to be minimized or maximized is given as a parametrised *objective function*: $f : D^n \to \mathbb{R}$. The *constraints* may be formulated as a set of functions and bounds: $C(x) \leq b$. The MST can then be formulated as follows: Given an undirected graph $G = (V, E)$ and a non-negative cost function $d : E \to \mathbb{R}^+$,

$$\text{minimize:} \quad C(T) = \sum_{\{i,j\} \in E_T} d_{ij} \tag{2.5}$$

subject to:

$$T = (V, E_T) \tag{2.6}$$

$$E_T \subseteq E \tag{2.7}$$

$$\exists i \in S, j \in V \smallsetminus S : \{i, j\} \in E_T \qquad \forall S \subsetneq V : S \neq \varnothing \tag{2.8}$$

The cost function (2.5) gives the cost of the tree as the sum of the costs for all included edges. The constraints ensure that the tree is a graph using the same nodes as the original graph (2.6), that only edges of the original graph are used (2.7), and that every part of the original graph is connected to the remaining graph using edges of the tree (2.8).

Although these constraints are not given directly in the form of inequalities $C(x) \leq b$, they can easily be transformed. An equation

$C(x) = b$ can be transformed using two inequations $C(x) \leq b$ and $C(x) \geq b$. A set M can be transformed by using a set of auxiliary variables $x_i \in \{0, 1\}$, and setting $x_i = 1$ for all members $i \in M$ and $x_i = 0$ for every other part of the domain D. The subset relation can then be formulated as a set of inequations $y_i \leq x_i$. And the existential quantification can be transformed into a sum: $\exists i \in M : p(i) \leftrightarrow \sum_{i \in M : p(i)} x_i > 0$. As such formulations may be harder to comprehend, we will try to use straightforward formulations for the problems in this thesis. However, in some cases a more formal definition is unavoidable.

A *problem instance* or simply *instance* is a realization of an optimization problem that gives values to all parameters. In the MST example, an instance would be a network, i. e. a set of nodes, a set of edges, and a cost function for these edges. The *problem size* n denotes the size of the problem instance, usually measured by the number of variables. In the network problems considered in this thesis, we will use the size of the network, i. e. the number of nodes in the network, as the problem size n. The number of variables used in the formulations may be larger, as each edge in a fully meshed network is usually represented by a variable, thus creating n^2 variables.

problem instance

problem size

Any set of assignments for all variables of a given problem instance that satisfies all constraints is called a *feasible solution* or simply a *solution* to this problem. In the case of the MST, such an assignment would represent a spanning tree for the given network. An assignment that violates one or more constraints of the given problem is called an *infeasible solution*. This is the case when the tree does not reach all nodes of the given network, or when edges are used that are not included in the original network. The *cost* or *utility* of a solution can be calculated using the objective function. If a solution is feasible and there are no feasible solutions with a lower cost, then this solution is called *optimal*. In the case of a maximization problem, it is optimal when no other feasible solution has a higher utility value.

solution

If the domain D of the variables of an optimization problem is discrete, the problem can be called a *combinatorial optimization problem*. Network optimization problems usually belong to this class. The counterpart is given by *continuous optimization problems*. A simple example is the minimization of a continuous function $f : \mathbb{R}^n \rightarrow \mathbb{R}$. In this thesis, only few continuous optimization problems will be mentioned, as the main problems considered in this thesis are combinatorial optimization problems.

combinatorial vs. continuous

formal
definition

Garey and Johnson give a more formal definition for combinatorial optimization problems in [57]. According to them, a *combinatorial optimization problem* is either a minimization or a maximization problem Π consisting of

- a set of instances D_Π

- a finite set of candidate solutions $S_\Pi(I)$ for each instance $I \in D_\Pi$

- and a function $m_\Pi(I, \sigma) \in \mathbb{Q}^+$, which assigns a solution value to all solutions $\sigma \in S_\Pi(I)$ of an instance $I \in D_\Pi$ of the problem Π.

The set of instances D_Π is described by a general description of the parameters of the problem Π. An instance $I \in D_\Pi$ is given by providing specific particular values for all these parameters. The set $S_\Pi(i)$ is defined by a statement of what properties a solution has to satisfy. The optimal solution for this problem Π is the solution $\sigma^* \in S_\Pi(I)$ which satisfies:

$$\forall \sigma \in S_\Pi(I) : m_\Pi(I, \sigma^*) \le m_\Pi(I, \sigma) \tag{2.9}$$

for a minimization problem, or

$$\forall \sigma \in S_\Pi(I) : m_\Pi(I, \sigma^*) \ge m_\Pi(I, \sigma) \tag{2.10}$$

for a maximization problem. In their definition, the set of candidate solutions needs to be finite, while other researchers also accept discrete solution sets [137]. Also, the restriction to rational positive value functions can be relaxed. If the solution values of a combinatorial optimization problem are irrational or negative numbers, the value function of this problem can still be mapped to a positive rational function, as the solution space is discrete or even finite.

naming
convention

When referring to a specific optimization problem, the same name as the name for the optimal solution can often be used, e. g. MST refers to the problem of finding such a tree as well as the optimal solution to this problem.

2.2.2   Complexity

An optimization problem can easily be transformed into a decision problem. Instead of asking for the minimal solution, a question is formulated, whether there is a feasible solution with a cost lower than a given bound. This allows to transfer some of the complexity analysis theory from decision problems to optimization problems.

One important concept is the $\mathcal{NP}$-completeness of a problem. The term of $\mathcal{NP}$-completeness does not apply to optimization problems unless they are transformed to a decision problem. Instead, the weaker term $\mathcal{NP}$-hardness is used. The class $\mathcal{NP}$ defines the set of decision problems that can be solved in polynomial time by a non-deterministic Turing machine. A problem is $\mathcal{NP}$-hard, if every problem in $\mathcal{NP}$ can be reduced to this problem in polynomial time. If it also belongs to the class $\mathcal{NP}$, it is called $\mathcal{NP}$-complete. When proving that a problem is $\mathcal{NP}$-hard or $\mathcal{NP}$-complete, the reduction from one known $\mathcal{NP}$-complete problem is sufficient.

$\mathcal{NP}$-hard

The first known $\mathcal{NP}$-complete problem was the Boolean Satisfiability Problem (SAT), proven by Stephen Cook in [29]. His proof reduced the functionality of a non-deterministic Turing machine to a Boolean Satisfiability Problem, thus reducing all $\mathcal{NP}$ problems to SAT. A very thorough analysis of $\mathcal{NP}$-completeness as well as a large list of $\mathcal{NP}$-complete problems and proofs is given by Garey and Johnson in [57].

The consequence for optimization problems that belong to the class of $\mathcal{NP}$-hard problems is that there is no polynomial time algorithm to solve them, unless $\mathcal{P} = \mathcal{NP}$. In this equation, $\mathcal{P}$ stands for the class of problems that can be computed in polynomial time by a deterministic Turing machine. If a polynomial time algorithm for any of the $\mathcal{NP}$-hard problems was found, all of the $\mathcal{NP}$-complete problems could be solved in polynomial time by reducing them to the solved problem. It is a common conjecture that $\mathcal{P} \neq \mathcal{NP}$ [57]. The network optimization problems considered in this thesis are $\mathcal{NP}$-hard, as will be shown in the respective chapters. The MST does not belong to this class, it can be solved in polynomial time using the algorithms of Prim [145] or Kruskal [94].

2.2.3   Solution strategies

The formulation for an optimization problem does not necessarily indicate a solution strategy for this problem. Instead, it often only defines feasible solutions and their costs. In order to solve an optimization problem, a suitable approach has to be found.

A feasible, but often impracticable, approach for combinatorial optimization problems with finite domains D is *exhaustive search*. By enumerating all assignments for the variables of the problem, the feasible

exhaustive search

solutions can be determined and the costs for these solutions can be calculated. This approach requires exponential time: $\mathcal{O}(|D|^n)$. This is assuming that the objective function and all constraints can be computed in exponential time.

branch and bound

A more efficient approach is *Branch and Bound* (B&B) [96]. In this approach, the solution space D^n is partitioned iteratively, and each partition that can be shown to contain only infeasible or sub-optimal solutions is disregarded. The efficiency of this approach relies on the ability to define and calculate lower and upper *bounds* for a given partition of the solution space. Only the bounds are compared, no solution has to be generated to determine that a certain partition of the solution space is not worth further investigation. If such bounds cannot be efficiently computed, then B&B reverts to an exhaustive search. In general, B&B still needs exponential time.

bounds

Bounds are also important outside of B&B. The cost of any feasible solution can serve as an *upper bound* for a minimization problem. It shows that the cost of the optimal solution is lower than this bound, or equal. However, bounds can often be calculated without the knowledge of a feasible solution or even without indicating a possible solution. One possibility for calculating *lower bounds* is to relax some of the constraints of the optimization problem. In the example of the MST, ignoring constraints (2.6) and (2.8) gives a simple lower bound of $l_0 = 0$. A more useful lower bound l_S can be found by sorting the edges by their costs and summing up the $n - 1$ cheapest edges. This will most likely violate constraint (2.8), as some of these edges may produce cycles.

gap and excess

When an upper bound u and a lower bound l is known, the *gap* $g = \frac{u-l}{l}$ can be calculated. The gap can be used as a quality measure for the algorithms that produced the bounds. The simple lower bound l_0 results in an extraordinarily high gap, indicating that this bound has only limited use. A similar measurement for minimization problems is the *excess*, i. e. the gap between the cost of a given solution and the optimal cost. The cost of an optimal solution has to be known for this calculation. The excess can then be used to evaluate and compare different algorithms for the same problem. If no optimal cost is known, the excess is sometimes measured as the gap to a best known solution or to a lower bound.

approximations

Solutions that are not optimal can still be useful. An *x-approximation* gives solutions with a cost that is at most x times as high as the cost of the optimal solution (or with a utility that is at least one x-th of the optimum in case of a maximization problem). Such approximations can sometimes be calculated much easier than the optimal solution. An approximation scheme that allows to calculate arbitrarily low $(1+\varepsilon)$-approximations in polynomial time is called a *Polynomial Time Approximation Scheme* (PTAS). The polynomial time complexity of the PTAS refers to the problem size n. The PTAS usually uses exponential time with respect to $1/\varepsilon$. Many optimization problems allow approximations or a PTAS, however, there are examples of optimization problems, where no such approximation can be found efficiently.

2.3 HEURISTICS

Many optimization problems are hard to solve with exact solvers. In these cases, heuristics can be used to find good solutions. This is especially true for $\mathcal{NP}$-hard optimization problems, such as the network optimization problems discussed in this thesis. The term heuristic derives from the Greek verb ευρισκειν, which means to find. Heuristics are algorithms for optimization problems based on some general strategy for searching through the solution space. They work generally well in practise, but they do not provide proof for the optimality of the generated solutions, and they usually do not provide a guarantee for the approximation ratio. Some researchers also include the lack of a quality or efficiency guarantee in their definition of heuristics. Juraj Hromkovič provides an even stricter definition in [72], which also excludes randomized algorithms that provide a bounded constant probability $p > 0$ for the optimality of the solutions.

In this thesis, we will use the term heuristics to denote a general solution technique for optimization problems that implements a certain strategy aiming at good quality solutions but does not provide a guarantee for the optimality of the generated solutions. This broad definition also includes approximation algorithms and improving heuristics that start from solutions generated by an approximation algorithm.   *definition*

The main advantage of heuristic approaches lies in their applicability to larger problem instances and in their ability to find good solutions in short time. This section gives a brief overview of well established heuristics, many of which will be applied to network optimization problems in the following chapters.

2.3.1   Evolutionary Approaches

Evolutionary Algorithms (EAs) form a class of heuristics that are inspired by biological evolution [9]. They are an effective means to tackle hard optimization problems. There are many forms of Evolutionary Algorithms, among them Genetic Algorithms (GAs) [40], Genetic Programming [34, 91], and Evolution Strategies [149]. Although these approaches differ in a number of details, they also show certain similarities, which can be used to define a general EA.

EA

An EA holds a *population* of one or more individuals, each representing an intermediate solution. The EA evaluates the fitness of these individuals using the cost function. It also searches for better solutions using the genetic operators mutation, recombination and selection. During *mutation*, an individual from the population is randomly changed.

mutation

Depending on the representation of the individual, this change may be applied by switching one or many bits, or by adding random values to the variables of the individual. *Recombination* merges genetical information from multiple parent individuals in one or more individuals. A

recombination

widely-used form of recombination is cross-over. Here, the representation of the parent individuals is split and the offspring is generated by taking one part of the solution from each parent.

In each generation, the EA uses mutation and recombination to create new offspring individuals. EAs without recombination exist, as well as EAs without mutation. Out of all generated offspring, only some are

selection

selected using a fitness criterion based on the problem's objective. Usually, the population is reduced back to the original size. However, there are approaches which allow the EA to adapt the population size. The selected individuals form the population for the next generation, and the process is continued. A simple EA without recombination and with a population of only one individual is shown in Figure 3.

There are two different approaches to selection. An *elitist* strategy preserves the best individual from the previous generation if no improvement was made. This ensures that the quality of the solutions is not reduced during the evolutionary search. An *extinctive* selection strategy ignores the previous generation even if no improvement was found. In this case, the best solution found during the search is stored, but the corresponding individual is deleted. This approach has some advantage, as it allows the EA to search through other parts of the solution space, and possibly also increases the diversity of the population.

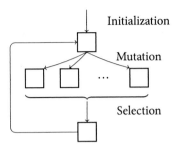

Figure 3: Flow diagram of a simple Evolutionary Algorithm (EA). In each generation, the individuals from the current population produce a number of offspring. The population is then reduced to their former size using some form of selection. In this simple EA only one individual and only mutation is used.

Ingo Rechenberg established a naming convention for the selection strategies of EAs in [149], where (μ, λ) refers to an EA with a population of μ individuals, from which λ offspring individuals are generated using mutation or recombination. The following selection reduces the population to μ again, selecting the μ best individuals from the offspring generation only. The $(\mu + \lambda)$-EA gives the preservative counterpart, where the best individuals are chosen from both the offspring and the previous generation.

EAs have been applied successfully to many optimization problems, such as the $\mathcal{NP}$-hard Travelling Salesman Problem (TSP) [174], or to find the optimal form of a flashing nozzle [149].

History

Evolutionary Algorithms are based on biological evolutionary theories formulated by Charles Darwin [36] and Gregor Mendel [108] in the 19th century. Both theories were combined into a modern evolutionary theory in the 1920s. These theories describe the evolution of species by evolutionary processes such as natural selection, mutation and crossing.

The theories have later been transferred from biology to mathematics and computer science. Some early evolutionary techniques have already been used in the late 1950s [10], but it was not until the 1970s

biology

EA

that EAs became very popular. There are various forms of EAs. The following paragraphs describe the four most prominent among them.

GA

John H. Holland introduced GAs, based on his work on cellular automata [69]. These GAs were used to solve optimization problems. Although the genetic operators are not different from the general operators described above, the GA differs from other EAs, as it uses a binary representation of the solution. This allows to restrict the genetic operations to bit switching.

EP

Evolutionary Programming was introduced by Lawrence J. Fogel as an approach to model artificial intelligence [10]. This type of EA evolves a finite state machine. Unlike Genetic Programming, it uses a fixed structure, where only the parameters for the finite state machine are evolved in an evolutionary approach.

GP

Genetic Programming was proposed by John R. Koza in [91]. This type of EA also evolves general computer programs to calculate arbitrary functions, however, they are modelled using a tree structure. The genetic operators are allowed to apply all kinds of changes to this tree structure. The tree defines a mathematical function, where leaf nodes are used either as constants or as input variables and interior nodes represent simple mathematical operations such as adding or multiplying values. Once a tree is built, the function can then be evaluated or tested against a given set of input-output pairs.

Genetic Programming and Evolutionary Programming are closely related. Techniques from one field have been transferred to the other field. Both have applications in fields usually attributed to artificial intelligence, such as prediction of unknown processes or classification problems.

ES

Evolution Strategies were introduced by Ingo Rechenberg in [149]. This type of EA again concentrates on optimization problems. An interesting speciality of Evolutionary Strategies is that they were used in conjunction with real world experiments to optimize real world phenomena, such as finding the optimal form of a flashing nozzle or of an aerodynamic wing.

2.3.2 Local Search Approaches

Local Search (LS) is a technique to improve a given solution by searching for better solutions in the neighbourhood of this solution. A local

neighbourhood

search is defined by its *neighbourhood* function. This neighbourhood function $N : D^n \to \mathcal{P}(D^n)$ defines for each solution $s \in D^n$ a set of

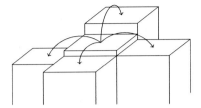

Figure 4: Local Search with a neighbourhood of four solutions. Each rectangle represents a solution, and the z-axis is used to indicate the cost of the solution. Starting from the initial solution shown in the middle, the move to a neighbouring solution may increase or decrease the cost. If the problem is a minimization problem, a best improvement local search will choose the solution to the left and continue from there.

neighbouring solutions $N(s)$. Usually, the neighbourhood function is modelled after a problem-specific set of moves. These moves are usually very simple, they may be designed to allow a quick calculation of the resulting objective. They form a neighbourhood that is a very small subset of the solution space.

A *Local Search* is an algorithm that starts from a given solution and continually moves from one solution to one of its neighbouring solutions, thus searching through the search space [120]. This search can also be seen as a walk through the neighbourhood graph. A very simple LS is *iterative hill climbing*, which only follows improving moves. If there is a better solution in the neighbourhood of the current solution, the LS is continued from this solution. Figure 4 illustrates this on a very abstract level. Here, neighbouring solutions are represented by neighbouring rectangles. Starting from the middle rectangle, the LS is presented with four neighbouring solutions. It will select one of these solutions and continue the search from there. In the following, we will first discuss LS that follows only improving moves, and then present examples for LS variants which also allow non-improving moves.

Even though the MST is not $\mathcal{NP}$-hard, it can still be used to give an example for a LS. Starting from a spanning tree, a LS for the MST could be designed to exchange an edge from the tree with an edge from the remaining graph. This LS should avoid to split the tree, as this would not represent a feasible solution. The costs for these moves can be calculated by the difference of the costs for the swapped edges.

LS definition

When selecting the solution from the neighbourhood that should be followed, the LS uses a pivoting rule: The LS can be designed to pick the first improving solution, or to search the whole neighbourhood and pick the best improving solution. Using the first improvement strategy usually leads to shorter running times, though the time complexity of the LS is not reduced. Following the best improvement, one expects to find better solutions [65].

first vs. best improvement

If an improvement is found, the LS continues from this solution. The LS terminates once the neighbourhood of the current solution contains no improving solution. This last solution is the result of the LS; it is not necessarily a global optimal solution for the problem, however it constitutes a *local optimum*. Due to its design, the simple LS cannot leave a local optimum. Various techniques have been proposed to allow the LS to escape from local optima. The following paragraphs describe some of them.

local optimum

Variable Neighbourhood Search

A simple way to escape from a local optimum is to change the neighbourhood, i. e. to change the type of moves made by the LS. Different neighbourhoods may produce different local optima. This technique is used in *Variable Neighbourhood Search* (VNS) as proposed by Nenad Mladenović and Pierre Hansen in [122]. They use a series of pre-selected neighbourhood structures N_k ($k = 1, \ldots, k_{max}$). The VNS usually starts from N_1 and applies the LS until a local optimum for this neighbourhood has been found. Once the local optimum is reached, the VNS switches to the next neighbourhood N_{i+1}, until the last neighbourhood $N_{k_{max}}$ was searched. If an improvement has been found in any of these neighbourhood searches, the VNS switches back to the first neighbourhood N_1. This type of VNS is also called Variable Neighbourhood Descent, as it is using a first-improvement strategy to descend through the different neighbourhoods N_k.

change the neighbourhood

Other variants of the VNS are also possible. A random sampling approach could restrict the search in the neighbourhood N_k to a number of randomly generated solutions. This number can also be a function of k. When switching back, the VNS can also use any previous neighbourhoods N_i ($i = 1, \ldots, k - 1$) and is not restricted to switch back to N_1. The switch to the next neighbourhood could also be executed using larger steps, such that the neighbourhood N_{k+s} is used if N_k did not yield improvements.

variants

A very fruitful approach is to stepwise increase size of the neighbourhood. This way, the first LS can quickly improve the solution to a certain degree. The next LS can then start from an already good solution and try to find more improvements. If the stronger LS would have been applied to the initial solution, the computation time would have risen, as the stronger LS has to search a larger neighbourhood. However, using the stronger LS from the beginning may also lead to another local optimum, as the weaker LS might be misled by initial improvements.

Mladenović and Hansen applied their VNS to the TSP in [122]. They started from tours generated by a heuristic called GENI (generalized insertion) and used a parametrised LS called US (unstringing and stringing) in their VNS. The VNS iteratively increases the parameter k used in the LS. Both, LS and construction heuristic were proposed by Michel Gendreau *et al.* in [58], where they were simply executed sequentially. Gendreau *et al.* called the resulting heuristic GENIUS. The construction heuristic GENI iteratively includes the least cost vertex into the tour. In the LS US, one vertex is removed from the tour (unstringing) and another vertex is included (stringing). The parameter k refers to the number of vertices that are considered for inclusion. In the LS US, only the k closest neighbours of the removed vertex are considered. Obviously, the neighbourhood size is proportional to k. Gendreau *et al.* used only $k = 1$ or $k = 2$, and Mladenović and Hansen improved upon these results using only slightly larger computation times with the VNS approach.

example application

Evolutionary Local Search

LS can also be combined with an EA, resulting in a *Memetic Algorithm* (MA) as proposed by Pablo Moscato in [126]. After each of the genetic operations in the EA, a LS is performed. As the genetic operations mutation and recombination are random searches, the LS adds a sense of direction to the EA.

MA

A simple form of an MA is *Evolutionary Local Search* (ELS). It combines a $(1 + \lambda)$-EA with LS as shown in Figure 5. The LS is used to improve the intermediate solutions produced by the EA. An early form of an ELS was presented by Günther R. Raidl and Ivana Ljubić in [146] for a graph problem called the edge-biconnectivity augmentation problem.

ELS

There is also a possible source of confusion concerning the term ELS, as Evolutionary Local Selection as proposed by Filippo Menczer *et al.*

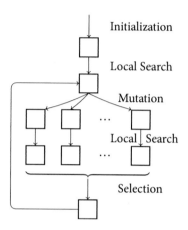

Figure 5: Flow diagram of an Evolutionary Local Search (ELS). Initialization, mutation and selection work just like in an Evolutionary Algorithm (EA). After each step of the EA, a local search is applied.

in [107] was also sometimes called ELS. The Evolutionary Local Selection algorithms are used for multi-objective optimization and are based on Simulated Annealing, which is described below. They introduce locality in the selection phase by establishing a scheme that allows a global selection based on local interactions between two individuals.

If the ELS is confined to produce only one offspring in each generation ($\lambda = 1$), it is equal to an *Iterated Local Search* (ILS) as introduced by Olivier Martin *et al.* in [105] and largely propagated by Thomas Stützle [169]. ILS was not originally motivated using evolutionary techniques, but rather as a means to overcome the problem of LS getting stuck in local optima. In the context of ILS, the mutation is called *perturbation* [71, 103].

The search pattern of an ILS is shown in Figure 6. When ILS is started, the LS quickly improves the initial solution and reaches a local optimum. Using perturbation, the ILS is able to escape the attraction basin of this local optimum and continue the search in other parts of the solution space. If the perturbation is too small, the same local optimum is found again. If the perturbation is too big, and the starting point for the following LS is not really related to the previous local optimum, the ILS resembles a random restart approach. In this case, the knowledge from the previous local search is lost.

ILS

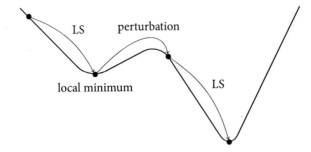

Figure 6: Search pattern of Iterated Local Search (ILS) on a minimization problem. The curve represents a 2d-projection of the objective function. The pattern also fits for Variable Neighbourhood Search (VNS) when Perturbation is replaced by a change of the neighbourhood.

The figure also applies to ELS or MAs. In an ELS, the mutation and the following LS are applied several times before switching to the best solution found in this generation. In an MA, the whole search is applied in parallel to multiple solutions, and the next generation is formed from the best solutions found in all these searches.

Tabu Search

Another way to overcome local optima in LS is *Tabu Search* (TS) [59]. During the LS certain moves are marked tabu, i. e. these moves are not applied and not evaluated. To this end, moves that would simply revert *mark as tabu* an applied move are stored in a tabu list. The main purpose of this tabu list is to prevent the search to cycle back to a former solution. Thus TS is allowed to accept worse solutions when all remaining moves are marked tabu.

One motivation behind this technique is the hope that the LS will eventually escape the attraction basin of the local optimum and find other local optima. The tabu list can also be used to speed up the search, as it reduces the size of the neighbourhood. The tabu list is usually limited in size, thus re-enabling forbidden moves after the search has progressed to another part of the solution space. The size of the tabu list should be chosen at least as large as the longest cycle that needs to be prevented, but larger tabu lists often result in better solutions [59]. Moves may also be removed from the tabu list when they are found to improve the solution.

don't look bits

A similar technique when considering graph optimization problems is given in the form of Don't Look Bits (DLBs) [14]. This technique was first applied to the TSP, where nodes that do not give an improving move are marked. They are ignored in the following steps of the local search. However, when they are part of an improving move, the mark is removed. This can happen, when an unmarked node is evaluated and the edge connecting these two nodes proves to be an improving edge. Using DLBs does not help to overcome local optima, but the computation time to reach a local optimum can be reduced considerably.

Simulated Annealing

The last approach mentioned in this section is *Simulated Annealing* (SA) [87, 22]. During SA, moves that increase the cost of the current solution are also allowed. The approach is modelled after the annealing procedure for heated metals. These materials form strong structures when they are cooled carefully, but a sudden cooling process usually results in fragile material. This is modelled in an SA using the notion of *temperature*

temperature. When the temperature is high, already formed local structures can be dissolved, i. e. worse solutions are also accepted. When the temperature is low, only improving moves are accepted.

In the SA, an unsuccessful move is accepted with a certain probability which depends on the current temperature and on the cost increase

accept bad moves

for this move. This probability is given by the Boltzmann-Gibbs distribution

$$p(E) = C \cdot e^{-E/T} \tag{2.11}$$

which gives the probability of finding a physical system at a given temperature T in a particular configuration with an energy of E. The constant C in this equation is a normalization constant. In an SA, the probability of accepting a non-improving move, i. e. a move that increases the cost by ΔE, is defined as

$$p(\Delta E) = C \cdot e^{-\Delta E/T} \tag{2.12}$$

Again, the constant C is used for normalization.

During an SA search, the temperature T is initially set to a high level and is reduced in the following steps. The temperature may also be increased again from time to time to dissolve the structures that have been build so far, in order to find other, possibly better structures.

When viewed as a form of LS, this is an effective technique to overcome local optima.

2.3.3 Ant Systems

The *Ant Colony Optimization* (ACO) approach replaces the undirected random mutation of EAs with a more sophisticated technique introduced by Marco Dorigo *et al.* in [48, 49]. The heuristic models the be-   *ACO* haviour of ants searching for food sources. Real ants leave a trail of *pheromones* wherever they go. Other ants follow these trails. When they arrive at a junction, they choose randomly based on the amount of pheromones on the trails. They more frequently choose stronger pheromone trails, but they also follow weak trails or explore new territories from time to time. Since pheromones dissipate over time and pheromone trails can be renewed more often on shorter paths than on longer paths, the ants will eventually find and follow the shortest path between food and nest.

The ACO heuristic models this behaviour by placing pheromones τ_i on building blocks i of the solution. These building blocks could be edges in a graph problem or a certain variable assignment in a general optimization problem. When generating solutions, the ACO randomly assembles these building blocks, and chooses based on the amount of pheromones τ_i and other heuristic information η_i using the following distribution:

$$P(i) = \frac{\tau_i \cdot \eta_i}{\sum_j \tau_j \cdot \eta_j} \qquad (2.13)$$

After each iteration the amount of pheromones τ_i on building blocks resulting in the best solution is increased, and other pheromones are reduced.

Another successful ACO variant was introduced by Thomas Stützle and Holger H. Hoos in [170]: The Max-Min Ant System (MMAS) is   *MMAS* an ant system characterized by having a lower and an upper bound $[\tau_{\min}, \tau_{\max}]$ on the pheromone trails to avoid the stagnation behaviour suffered by other ant systems. This stagnation behaviour is a result of reinforcing some edges by laying lots of pheromone trails along them while other edges are never used. This behaviour leads to a repeated selection of the same solution while other regions of the search space are not explored.

model limitations

These models abstract from some aspects of ant behaviour. Real ants use pheromones as a general form of communication. They are able to encode the path taken when searching for food differently than the path taken when the food source has been found. Ants are also often divided in groups working on special tasks, such as collecting food, building or cleaning the nest, or defending the nest against enemies. If there is an acute need for ants for a special task, other ants can be called using pheromones. An individual ant can switch its role, if such need arises. In the case of an attack, pheromones can also be used to stimulate aggressive behaviour.

applications

ACO algorithms of various forms have been used to solve different $\mathcal{NP}$-hard problems, such as the Travelling Salesman Problem (TSP) [170] or the Quadratic Assignment Problem (QAP) [43]. A comprehensive and very detailed overview of ACO and its applications is given in the book of Dorigo and Stützle [47].

ACO + LS

The ACO heuristics often already lead to good results, however, ACO can also be enhanced with other techniques. One example is to use LS, i. e. executing a LS from each randomly generated solution and using the improved solutions as a base for the next generation's pheromones.

2.3.4   Nested Partitioning

NP

A completely different kind of heuristic for global optimization problems is the *Nested Partitioning* (NP) approach introduced by Leyuan Shi and Sigurður Ólafsson in [155]. In this approach, the solution is constructed piece by piece, similar to ACO or any construction heuristic. In each step of the NP, the solution space D^n of the optimization problem is systematically divided. Each partition is then sampled, i. e. random solutions in that part of the solution space are generated and evaluated. The process is continued on the partition that contains the best solution. Thus, the NP dives deeper into the solution space. NP terminates when the best partition contains only one solution. An example of an NP is shown in Figure 7.

The partitioning of the solution space is usually accomplished by assigning values to one of the n variables of the problem or by splitting the range of one variable. The sampling can then be done by randomly assigning values to the not yet fixed variables, or by using a randomized construction heuristic for the given problem.

backtracking

NP can also backtrack. This is triggered by sampling the *surrounding region*, i. e. the part of the solution space not considered in the cur-

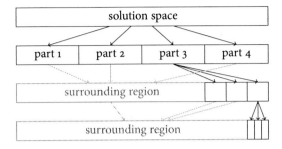

Figure 7: Work flow of a simple Nested Partitioning

rent iteration or depth of the NP. At depth 1, the surrounding region is empty. In each iteration of the NP, when the promising region is split to create the subregions for the next depth, the remaining regions are merged into the surrounding region. Samples for the surrounding region are also generated, using the same sampling methods as for the subregions. However, the NP can be set to generate less samples for the surrounding region, or use any other more sophisticated sampling scheme in order to intensify the search in the promising parts of the solution space. This will often reduce the ability of the NP to recognise an earlier mistake, though.

NP is a generic method for optimization. Just like all other heuristics described in this section, it can be applied to various kinds of optimization problems. The only assumption made by NP is that the solution space of the optimization problem is finite. NP does not need an initial solution like LS. NP can further be combined with other heuristics, e. g. the samples can be generated by an ACO-like approach, or they can be improved by LS. NP can also be parallelized, as the sampling procedure for different subregions or even for the different samples can be executed independently. *advantages*

NP algorithms have been successfully used for a number of $\mathcal{NP}$-hard optimization problems such as the Travelling Salesman Problem (TSP) [157], product design [158] and scheduling parallel machines with flexible resources [134]. Local search heuristics have been implemented in an NP algorithm, such as using 2-exchange or 3-exchange local search in case of the TSP [157]. NP was also used in combination with a number of other heuristic algorithms, such as Genetic Algorithms [158] or Max-Min Ant Systems [159]. *applications*

2.4   DISTRIBUTED ALGORITHMS

The optimization problems discussed in this thesis can be solved using global view optimization. However, in many applications, this is impracticable, as the knowledge about the whole network is vastly distributed. This thesis presents distributed algorithms to approach the optimization problems. To understand the functionality of these algorithms, two major concepts are needed: distributed systems and self-organization.

2.4.1   *Distributed Systems*

A *distributed system* can be defined as an interconnected collection of autonomous computers [173]. The participating computers are also called *nodes* in this distributed system. These nodes are autonomous, meaning that they are equipped with their own control and their own memory. The connection between these nodes allows them to exchange information.

There are other definitions that put more emphasis on certain aspects of distributed systems. Andrew S. Tanenbaum argues that the autonomous nodes should be transparent to the users of a distributed system [172]. In his definition, the collection of independent computers should appear as a single coherent system. George F. Coulouris *et al.* focus on loosely-coupled distributed systems and restrict their definition to networked computers [31].

In this thesis, we use the term *distributed system* to denote a system of multiple nodes with sometimes limited computation power that are loosely connected and work together serving some common purpose. *definition* Using this broad definition also allows us to view a network of wireless nodes with very limited computational power and memory as a distributed system.

When talking about computer networks, a participating computer *computer* is referred to as a *node*. A node may *leave* or *join* the network. The *networks* process of nodes leaving or joining a network is called *churn*. Churn is often a problem for distributed algorithms as the data stored at the leaving nodes is often lost, and the new nodes have to be integrated in the network. Nodes are connected to other nodes via *links*. If a node is not directly connected to another node, they can often still communicate by having intermediate nodes relay their messages. Such connections are called a *multi-hop* connections. If there are two nodes in the network that are not connected via links or multi-hop connections, the network is *split* or *partitioned*. Each partition of the network can still function on its own. However, a *self-healing* system is expected to reconnect the network over time.

2.4.2 *Distributed Algorithms*

A *distributed algorithm* is an algorithm that runs on a distributed system. Using the definition of Garey and Johnson from [57], a distributed algorithm can be defined as a general step-by-step procedures for solving problems, that is executed on multiple independent but interconnected machines at once. It consists of local rules, uses messages to coordinate the nodes, and aims at providing a global function. Distributed algorithms differ largely from centralized algorithms. Main characteristics of distributed algorithms are their lack of knowledge of a global state, their lack of a global time-frame and some amount of non-determinism [173].

distributed algorithm

Each node has its own *local view*, as Figure 8 demonstrates. Knowledge from other parts of the system must be transferred via *messages*, and thus is potentially outdated. The local view of two entities can also be contradicting. In a centralized algorithm, the global state can be perceived by looking at each variable one by one. During this reading operation, a centralized algorithm does not need to expect that the data is modified. In a distributed algorithm, data that is stored on another node can be modified by each entity of the distributed system. Thus, decisions based on a global state cannot be made, the nodes in a distributed system have to decide based on their local view only.

no knowledge of global state

In a centralized system, a global time-frame can easily be established. As there is no concurrency between the processes of a centralized algorithm, these processes are totally ordered. A new process is started only after the old process has finished its operations. In a distributed system, this is not the case. Here, multiple processes can be executed at the same time. They are not allowed to change the same variables at the same time, but they can change different parts of the global state simultaneously. The individual actions of the distributed system can only be partially ordered. Also, the perception of the order of events may be different for different nodes in the distributed system. If an event is signalled by sending a broadcast message to all other nodes in the system, the message delay may result in the early reception of a signal

no global time-frame

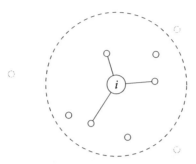

Figure 8: Local view of a node i in a distributed system. The node maintains links to direct neighbours in the given topology. It also knows some of the remaining nodes, but the major part of the network is unknown to the node, especially the links between remote nodes.

for a later event, if this event occurred closer to the receiving node in terms of message delay.

Since the nodes in a distributed system do not have knowledge of a global state and may perceive a different order of events, the whole distributed system can only be described by a number of concurrent processes. Even though such systems could still be analysed by complex models, such as Abstract State Machines [64], the differences in execution speed of the distributed nodes, possible unpredictability of message delays, or the interaction with the system and the environment may lead to the perception that the distributed system is largely characterized by non-determinism.

non-determinism

2.4.3 Self-Organization

The term self-organization was introduced by W. Ross Ashby in [8], to denote a system that is strictly determinate yet shows self-induced changes of its internal organization. He motivates this definition from the human nervous system, which can be modelled as a determinate physico-chemical system.

psychology

The term has since been applied to many sciences. Scott Camazine *et al.* give a definition of self-organization in biological systems [19]: "Self-organization is a process in which pattern at the global level of a system emerges solely from numerous interactions among the lower-level components of the system. Moreover, the rules specifying inter-

biology

actions among the system's components are executed using only local information, without reference to the global pattern." They give an example of emerging skin patterns of animals like the zebra or the giraffe, which are results of very simple bio-chemical rules.

Self-organization also occurs in other biological systems, such as ant colonies, in economical systems, such as the stock market, in physics, in chemistry, and in many other fields [68]. They all have in common, *other fields* that local interactions create global structures or behaviour. This emerging behaviour may be unwanted, e.g. a mass panic, but it can also be the main goal of a distributed algorithm.

In computer science, a *self-organizing system* is a system of many entities that serves a certain function without a central entity. It is usually characterized by showing some kind of structure on a global scale *computer science* that emerges from a very limited set of local rules. In [41], Hermann de Meer and Christian Koppen characterize self-organizing systems: The self-organizing system operates independently of its environment, and the boundaries between the system and the environment is defined by the system itself. An example is the membership in a P2P system. The self-organizing system is independent of identity or structure, and thus gains flexibility and adaptivity. In the P2P system example, each node could be replaced by another node joining the system, yet the self-organizing system would remain intact. The self-organizing system tries to maintain itself. It uses communication between the components which leads to feedback. It shows criticality and emergence, i.e. small local changes can have large global effects. It often reduces complexity, as an orderly system can be described using fewer rules than a fully chaotic system. De Meer even places swarm intelligence, ant systems and cellular automata in the context of self-organization.

Self-organization has become a very popular term, and subcategories of this term are also used to describe properties of distributed systems, such as *self-management, self-stabilization, self-configuration*, or *self-protection*. As the distributed algorithms developed in this thesis are used for optimization, the term *self-optimization* can be applied. Self-optimization can be seen as a combination of optimization and *self-* self-organization. *optimization*

MINIMUM ROUTING COST SPANNING TREES

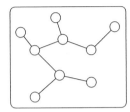

3.1 MOTIVATION

One aim in Peer-to-Peer (P2P) topology construction is to minimize the load on the peers. If load is measured by the number of links each peer has to maintain, then minimizing the total number of links in the network is an important goal. This can be achieved by using a spanning tree topology.

What's in a tree?

An additional advantage of this topology is revealed when it is used for broadcasts. Using tree topologies allows a very simple forwarding scheme: Each node forwards the broadcast message to all other neighbours. Since the tree topology is cycle-free, the broadcast message will be delivered to every node only once. Using any other kind of topology brings additional administrative overhead. The nodes need to avoid sending out duplicates of the broadcast message, as this would lead to an unlimited number of messages sent, caused by a single broadcast [24]. Structured P2P systems, such as Chord [167] or Pastry [151], also use trees for broadcasts. However, they need to add a range of recipients to the forwarded broadcast messages, to ensure that no cycle is created.

There are also disadvantages to tree topologies: The tree is not robust against edge or link failures. If a link fails, the network will be split. A practical P2P system based on a tree topology should incorporate some sort of repair mechanism to deal with these network partitions.

A second, but also important aim when constructing P2P topologies is the minimization of the message delay. As the P2P topology essen-

delay minimization

tially forms an overlay network which uses the underlying structures, each link in the P2P overlay may be built from a number of connections in the real network. These links are usually also used by other connections besides those of the P2P system. They may suffer congestions, and they are affected by routing decisions. In an underlay-aware P2P system, links that are heavily congested or links that use a large number of connections in the real network should be avoided. Such links can be recognized by their Round Trip Time (RTT) or by other link parameters, such as packet loss or bandwidth. So, trying to find a topology that minimizes the RTTs on each link helps to reduce the load on the underlying network. Many P2P applications, such as desktop grids, multimedia streaming or even content sharing, also directly benefit from a reduced message delay between the peers.

When combining both goals, minimizing the number of links and minimizing the message delay between the peers, the problem arises to find a common spanning tree that minimizes the routing cost for all pairs of nodes in the network. In this chapter, we present algorithms for constructing such topologies. The chapter is structured as follows: *structure* Section 3.2 provides the necessary mathematical background. In Section 3.3, we present a global view heuristic for finding such topologies, and Section 3.4 shows how delay-minimal tree topologies can be constructed and maintained in a self-organizing fashion. Section 3.5 summarizes the findings of this chapter.

3.2   MATHEMATICAL BACKGROUND

The problem of finding a spanning tree that minimizes the the sum of all distances is an optimization problem called Minimum Routing Cost Spanning Tree (MRCST). The MRCST is a special case of the Optimum Communication Spanning Tree (OCST), which was introduced by T. C. Hu in [73]. Both the OCST and the MRCST are $\mathcal{NP}$-hard [81]. They are listed as problem ND7 in [57].

first formulation Both problems can be formulated as follows: Given a weighted graph $G = (V, E, d)$ and a demand matrix r, find a spanning tree $T \subseteq E$ that minimizes the routing cost:

$$C(T) = \sum_{u \in V} \sum_{v \in V} d_T(u, v) \cdot r(u, v) \qquad (3.1)$$

The function $d_T(u, v)$ denotes the length of the unique path from u to v in T, i.e. the sum of the distances $d_{i,j}$ of all edges of this path.

In the MRCST, the demand is constant $r \equiv 1$, while in the OCST it is represented by a full demand matrix $r : V \times V \to \mathbb{R}$.

Using a directed tree rooted at an arbitrary node $v^* \in V$ in combination with the parent representation for trees as introduced in Section 2.1.2, equation (3.1) for the MRCST can be reformulated to show the weights on the edges:

second formulation

$$C(T) = \sum_{v \in V \setminus \{v^*\}} C_T(v) \qquad (3.2)$$

$$\text{with } C_T(v) = 2 \cdot s_T(v) \cdot (n - s_T(v)) \cdot d\left(v, p_T(v)\right) \qquad (3.3)$$

Here, $n = |V|$ denotes the size of the tree, $s_T(v)$ denotes the size of the subtree rooted at v, $p_T(v)$ denotes the parent of v in the directed tree T, and $(v, p_T(v))$ denotes the edge between v and its parent. The cost $C_T(v)$ of this edge is calculated by weighting the edge's length with the number of connections that traverse it. Since each node needs to communicate to every other node ($r \equiv 1$), the number of connections traversing an edge is twice the product of the number of nodes on one side of the edge times the number of nodes on the other side of the edge. The root node v^* does not contribute to the cost, it has no parent.

Using formulation (3.2), we can calculate the routing cost of a tree T in linear time, compared to a quadratic time for formulation (3.1).

This formulation also shows that an edge in the centre of the tree – especially an edge that splits the tree exactly in two halves – has a higher weight than an edge connecting a leaf to the tree.

edges in the centre have a higher weight

3.2.1 *Bounds and approximations*

Every spanning tree can serve as an upper bound for the MRCST. However, Bang Ye Wu and Kun-Mao Chao showed that the Shortest Path Tree (SPT) rooted at the median node $v^* \in V$ is already a 2-approximation of the MRCST [188]. This median node v^* is the node that has the minimal distance to all other nodes in the network when distance is measured over shortest paths only. This node can be found in polynomial time by constructing all SPTs, and picking the shortest one. The SPTs can be constructed in polynomial time using Dijkstra's algorithm [44] or the algorithm of Bellman and Ford [13, 55].

bounds

This approximation can also be used to calculate a lower bound. The solution for the MRCST must have a cost at least half as high as the 2-approximation. There are other lower bounds, e. g. the All-Pairs-Shortest-Path (APSP) lower bound. If the weight function of the graph obeys the

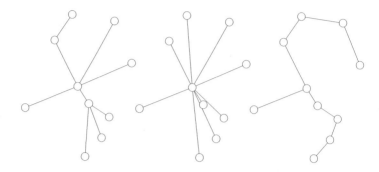

Figure 9: Example trees in the Euclidean plane. From left to right: Minimum
Routing Cost Spanning Tree (MRCST), Shortest Path Tree (SPT)
rooted at the median, and Minimum Spanning Tree (MST)

graph
properties
triangle inequality, the SPT will be a star, and the APSP lower bound
will be the same as the sum of all distances.

Based on the 2-approximation Wu *et al.* proposed a Polynomial Time
Approximation Scheme (PTAS) [189, 190]. Using this PTAS, a $(1 + \epsilon)$-
approximation can be found in $\mathcal{O}(n^{2\lceil \frac{2}{\epsilon} \rceil - 2})$ time. They also show how
the general case can be reduced to the metric case by replacing all edges
that violate the triangle inequality with their corresponding shortest
path. A solution for this metric closure graph can be transformed back
to a solution for the original graph without increasing its cost.

3.2.2   Relation to Other Graph Problems

Figure 9 gives an example for the MRCST, comparing it to an SPT and
the Minimum Spanning Tree (MST) for the same random Euclidean
MRCST vs.
SPT and MST
network. The shown SPT is routed at the median node in this network.
It therefore represents a 2-approximation for the MRCST. Since Euclidean distances were used, the triangle inequality holds, thus the direct edge is always a shortest path and the SPT degenerates to a star.
However, with only a few modifications, this SPT can be transformed
into the MRCST for this instance. The same holds for other instances
as well, including instances with distances that violate the triangle inequality. In fact, both SPT and MRCST do not select edges that violate
the triangle inequality.

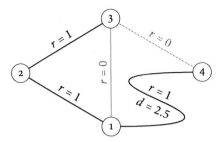

Figure 10: Example network for the OCST. In this example, each edge has a
length of $d = 1$, except for edge (1,4). The demands r are also shown
on the edges, and no demand that needs to be routed over multiple
edges exists in this example. Edge (1,4) violates the triangle inequality, however, it is used in the OCST for this network, which is highlighted in the graph.

The MST can be seen as a simpler form of the MRCST, as it minimizes the sum of the lengths of the tree edges instead as the sum of
the lengths of all paths. It can be found in polynomial time using the
algorithms of Kruskal or Prim [94, 145].

The OCST differs from the MRCST, as it allows triangle inequality violations. An example is shown in Figure 10. In this example, almost all *MRCST vs.*
edges have a length of $d = 1$. The demand matrix is sparse, containing *OCST*
only three pairs of nodes. In the OCST for this example, the demand
between nodes 1 and 4 is routed over the direct link, although this link
violates the triangle inequality. If the shortest path (1,3,4) was used for
this connection, the demand between nodes 1, 2 and 3 would have to
be rerouted, increasing the total communication cost. The MRCST for
this network is a star from node 3.

3.2.3   Applications in Peer-to-Peer Topology Construction

In a P2P network, several design goals can be defined: The P2P system
might try to minimize the total network load. If one assumes that a *minimize*
longer RTT on an overlay link correlates with a higher number of hops *network load*
in the underlying network, then such links should be avoided. Trying
to find paths with shorter RTT helps reducing the load on the underlying network.

minimize RTT

If the objective of the P2P system is to minimize the response time for frequent queries, this directly translates to a minimization for the RTT.

desktop grid

A P2P system can also be used in a desktop grid. Here, the nodes contribute spare computational power for other nodes to use. Any participant in a desktop grid can submit a job to the system which is then contributed to as many nodes as necessary. Regardless of the kind of job, a valuable design goal is to minimize the time it takes for the job to start, i. e. the time until a sufficient number of peers in the system has responded to the initiator indicating their willingness to calculate the job. Minimizing the average response time in the P2P system also helps in this scenario. For some jobs a certain level of cooperations between the workers on each sub-task is also required. In this case, the average RTT between the workers should be minimized. All this can be achieved by the MRCST.

3.3 GLOBAL OPTIMIZATION

In this section we present a global view heuristic for the MRCST. This section is structured as follows: Section 3.3.1 gives an overview of established exact as well as heuristic algorithms for the MRCST. Our heuristic is described in Section 3.3.2. We also provide test instances based on real world Internet measurements in Section 3.3.3, and present results of experiments on these instances as well as results from a fitness landscape analysis in Section 3.3.4. Section 3.3.5 summarizes our findings for the global view optimization of the MRCST.

3.3.1 *Related Work*

introduction

Since the MRCST was introduced as a special case of the OCST by T. C. Hu in [73], it has received much attention from researchers. In his paper, Hu motivated the problem by an application in telecommunication networks, and called the MRCST variant *optimum distance spanning tree*. The name Minimum Routing Cost Spanning Tree (MRCST) already appears in early papers such as [187], until then it was often approached as just another case of the Network Design Problem [81]. In some papers, the MRCST is also called *minimum communication spanning tree* [139] or *minimum routing cost tree* [21].

The MRCST was soon proven to be $\mathcal{NP}$-hard by David S. Johnson *et al.* [81, 57], by equivalence to the Simple Network Design Problem $\mathcal{NP}$-*hard* (SDNP), which was shown to be $\mathcal{NP}$-hard by reduction from the Exact 3-Cover problem (X3C, listed as problem SP2 in [57]).

In [1], Ravindra K. Ahuja and V. V. S. Murty presented a construc- *AMLS* tion heuristic and a local search for the OCST. They introduced steps to reduce the time complexity for these algorithms from $\mathcal{O}(n^5)$ in a naïve approach to $\mathcal{O}(n^3)$. Details of this local search AMLS and an adaptation of AMLS for the MRCST can be found in Section 3.3.2 on page 42.

In [82], Bryan A. Julstrom presented two Genetic Algorithms (GAs) and a simple stochastic hill-climber for the MRCST, and compared two *GAs* different codings for spanning trees. Although the GAs are less effect-ive than the simple stochastic hill-climber, Julstrom showed that using a coding scheme based on Prüfer numbers gives better results than an edge-set representation of the spanning tree.

Alok Singh presented a perturbation based local search in [162] us-ing a randomized variant of the AMLS. However, he failed to make use *perturbation* of the reduced complexity Ahuja and Murty presented in [1], so his *based LS* local search is weaker, but has the same high time complexity as the AMLS ($\mathcal{O}(n^3)$). Using the same test instances as Julstrom, he showed that local search leads to better results than the simple stochastic hill-climber.

Matteo Fischetti *et al.* presented two exact algorithms in [54]. Their *exact* most effective algorithm is based on Branch and Bound (B&B) and uses *algorithms* an integer programming formulation based on a path-representation. Although this formulation uses an exponential number of variables, it is preferred over a flow formulation with $\mathcal{O}(n^4)$ variables. Using this algorithm, the authors can solve instances of up to 30 nodes, and man-age to find optimal solutions for some of their instances with 50 nodes. This shows that trying to find optimal solutions is still infeasible for larger instances.

The fastest heuristic for the MRCST was presented by Rui Campos and Manuel Ricardo in [21]. Their algorithm iteratively adds vertices to *fast* the current tree. As a selection criterion, it uses a compromise between *construction* the shortest edge and the vertex with the highest degree, i. e. the highest number of adjacent edges leading to not yet included nodes. Using only shortest edges would result in a MST, while considering the degree of the included vertex helps to reduce the diameter of the resulting tree. This algorithm produces trees with similar costs as the best SPT. The

$s \leftarrow$ INITIALIZATION$()$
$s \leftarrow$ LOCALSEARCH(s)
mut $\leftarrow n$ *set mutation rate to problem size*
for *iter* = 1 ... κ **do**
> $s^* \leftarrow s$
> **for** i = 1 ... λ **do**
>> **if** *mut* = n **then**
>>> $s_{temp} \leftarrow$ INITIALIZATION$()$
>>
>> **else**
>>> $s_{temp} \leftarrow$ MUTATE(s, mut)
>>
>> $s_{temp} \leftarrow$ LOCALSEARCH(s_{temp})
>> $s^* \leftarrow \min\{s^*, s_{temp}\}$
>
> **if** $s^* < s$ **then**
>> $s \leftarrow s^*$
>
> **else**
>> mut $\leftarrow$ round$(\text{mut} \cdot \alpha)$ *decrease mutation rate*

return s

Figure 11: The ELS Framework

advantage of this approach becomes clear when it is applied in sparse networks. In such networks, it can be computed faster than calculating all n SPTs and picking the best one.

3.3.2 *Evolutionary Local Search*

The heuristic described in this section is based on Evolutionary Algorithms (EAs) and local search, and can be defined as an Evolutionary Local Search (ELS). The ELS works by starting from a valid tree and applying small changes, called moves, to the current tree. In order to escape from local optima, a mutation step is applied. The ELS framework is similar to Iterated Local Search (ILS) [103] or Memetic Algorithms (MAs) [126]. Its structure is shown in Figure 11. It differs from a pure ILS by using more than one offspring in each iteration. It also differs from usual MAs, because it does not use recombination.

In the following sections, the different parts of the ELS are explained.

Initialization

The first tree in the ELS is produced by connecting all nodes in random order to a random node that is already part of the tree. By starting from a random tree, we give away the opportunity to use an approximation as a starting point, and therefore cannot guarantee an approximation ratio for the ELS. However, random trees can be generated faster than any approximation, and they give a higher diversification, allowing the ELS to explore different parts of the solution space.

start from random trees

Mutation

In order to adapt the mutation rate to the current phase of the search, we use the following scheme: We use a higher mutation rate in the beginning of a run and decrease the mutation rate after each non-improving step. The ELS stays at the highest mutation rate as long as this gives an improvement. Thus it can fall back to random restart. When the search reaches better solutions, and random restart fails to find further improvements, the ELS gradually shifts from exploration towards exploitation using the mutation rate adaptation scheme.

adaptive mutation rate

The actual mutation is done by moving random subtrees to random new positions. However, in the beginning of the search, when the mutation rate demands to move n subtrees to new positions, the ELS simply generates a new random tree instead. The mutation rate is reduced by 20 % in each non-improving iteration ($\alpha = 0.8$ in Figure 11), and to prevent the ELS from being stuck once the mutation rate drops too low, the lowest mutation rate is set to 2.

Population and Selection

The ELS uses a population of only one individual. In each generation, it produces λ offspring using mutation and local search, out of which the best individual replaces the original if there was an improvement. Thus, when only $\lambda = 1$ offspring is produced, the ELS behaves similarly to an ILS with $\kappa + 1$ iterations, but still adapts the mutation rate to the current search progress. With $\kappa = 1$, the ELS equals random restart with $\lambda + 1$ iterations. Thus by varying the parameters κ and λ, we can observe how well a specific local search is suited for random restart, ILS, and ELS.

one individual only

The ELS can also be expanded to a full Memetic Algorithm (MA) by using a larger population and a suitable recombination.

Local Search

Since the MRCST is a special type of an OCST, each local search for the OCST can also be used for the MRCST. The most prominent local search for the OCST is the Ahuja-Murty Local Search (AMLS) [1]. But any local search for trees can be used as well. In this section, we will show the local searches used in our ELS, starting with the strongest, but most expensive one.

AHUJA-MURTY SEARCH   Ravindra K. Ahuja and V. V. S. Murty presented a construction heuristic as well as a local search in [1]. Both algorithms have the same structure, as they calculate the cost for an edge that is inserted in the tree. In the AMLS, the tree is split in two parts by removing an edge e. Both parts, called S and $\overline{S}$ are reconnected using the least expensive edge $\{i, j\}, i \in S, j \in \overline{S}$. Two sets of variables, w_i and h_i, are calculated to determine the cost for an edge. The total demand from node i to the other part of the tree is stored in w_i, while h_i gives the cost for routing the demands of the whole tree part to node i:

AMLS

$$w_i = \begin{cases} \sum_{j \in \overline{S}} r(i, j) & i \in S \\ \sum_{j \in S} r(i, j) & i \in \overline{S} \end{cases} \tag{3.4}$$

$$h_i = \begin{cases} \sum_{j \in S} w_j \cdot d_T(i, j) & i \in S \\ \sum_{j \in \overline{S}} w_j \cdot d_T(i, j) & i \in \overline{S} \end{cases} \tag{3.5}$$

Here, $d_T(i, j)$ is the length of the path from i to j in the corresponding part of the tree. Figure 12 illustrates this calculation. Using these values, the increase in cost can be computed as the sum of the routing cost to i and j in their respective parts of the tree plus the cost for routing the demands over the new edge $\{i, j\}$:

$$\alpha_{ij} = h_i + h_j + d_{ij} \cdot \sum_{x \in S} w_x \tag{3.6}$$

By removing the initial edge, the cost is reduced according to the same calculations. The total gain g of exchanging edge e with $\{i, j\}$ is simply:

$$g = C(T) - C(T \cup \{\{i, j\}\} \smallsetminus \{e\}) = \alpha_e - \alpha_{ij} \tag{3.7}$$

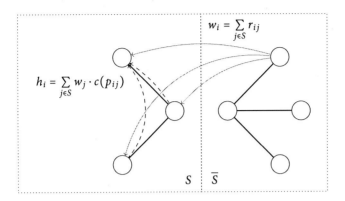

Figure 12: Ahuja-Murty Local Search. After the tree is split in S and $\overline{S}$, the values for each w_i and h_i are calculated. To repair the tree, an edge connecting both parts has to be inserted. Its cost can be calculated using w_i and h_i.

Since the MRCST uses equal demands $r \equiv 1$, the calculation can be simplified. The demand only depends on the size of the two parts of the tree:

$$w_i = \begin{cases} |\overline{S}| & i \in S \\ |S| & i \in \overline{S} \end{cases} \tag{3.8}$$

$$h_i = \begin{cases} w_i \cdot \sum_{j \in S} d_T(i,j) & i \in S \\ w_i \cdot \sum_{j \in \overline{S}} d_T(i,j) & i \in \overline{S} \end{cases} \tag{3.9}$$

$$\alpha_{ij} = h_i + h_j + d_{ij} \cdot w_i \cdot (n - w_i) \tag{3.10}$$

The first step in this calculation lies in determining the two parts of the tree and determining the path lengths. All path lengths in the tree can be calculated in $\mathcal{O}(n^2)$ time. They can be stored, as they do not change until the tree is changed. Determining the side each node is on can be done in $\mathcal{O}(n)$ time. This has to be done for every edge that is to be removed. Calculating all h_i values can be done in $\mathcal{O}(n^2)$ time, and again they have to be recalculated for every edge to be removed. So calculating the gain for every new edge can be done in $\mathcal{O}(n^2)$ time, and finding the best pair of edges to be exchanged can be done in $\mathcal{O}(n^3)$

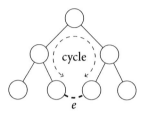

Figure 13: Introducing a cycle in a tree by inserting an edge e. To repair the tree, an edge from the cycle has to be removed.

time. The time complexity of the AMLS is the same for MRCST and OCST.

If the savings α_e for the removed edge e are less than the best gain found so far, the remaining h_i and w_i values don't need to be calculated. This simple speed-up does not change the complexity, though.

CYCLE SEARCH   During the AMLS, the tree is split first, before reconnecting its two parts by inserting an appropriate edge. This can be reversed: Inserting an edge in a spanning tree creates a cycle. To repair the tree, the cycle has to be broken by removing an edge from the cycle. The resulting graph is again a spanning tree.

CYCopt   In CYCopt, each of the $\mathcal{O}(n^2)$ edges is tried to be inserted. The insertion creates a cycle of at most n edges. If the tree is balanced, the expected cycle length is in $\mathcal{O}(\log n)$. Figure 13 illustrates this. For the following explanation, we enumerate the nodes in the cycle $(1, 2, 3, \ldots, m, 1)$ and call the inserted edge $(1, m)$.

Determining the gain of removing an edge from the cycle can be done in $\mathcal{O}(1)$ time, using a set of auxiliary variables. First, the number of paths crossing the cut $(i, i + 1)$ can be determined as $2 \cdot \left(\sum_{x=1}^{i} c_x \right) \cdot \left(\sum_{x=i+1}^{m} c_x \right)$. Here, m denotes the number of edges in the cycle, and c_x gives the number of nodes at that point in the cycle, i. e. node x and its subtree. The factor 2 is needed because paths exist for both directions. See Figure 14 for an example with $m = 5$ nodes in the cycle. In this example, $c_1 = c_3 = c_5 = 1$, $c_2 = 2$, and $c_4 = 3$. The cut at edge $(2,3)$ affects $2 \cdot (1 + 2) \cdot (1 + 3 + 1)$ paths. Calculating the gain on the cut edge can be done by multiplying this value with the length of the cut edge $d_{i,i+1}$. Using running sums, this calculation can be done in $\mathcal{O}(1)$.

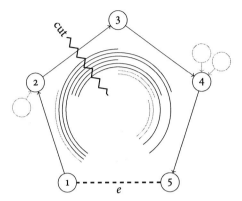

Figure 14: The gain of removing an edge from the cycle. By inserting edge $e =$ (1,5), a cycle (1,2,3,4,5,1) is formed. Cutting an edge, e. g. (2,3), affects some of the paths between those nodes, here: {1,2}×{3,4,5}. After the cut, these paths are routed over edge e, taking the opposite direction in the cycle, but the remaining paths are left untouched.

However, this is not the total gain of this exchange, as the paths extend before and after the cut edge. Since these paths have different lengths and are used by different numbers of nodes c_x, the calculation seems expensive at first glance. Nevertheless, it can still be achieved in $\mathcal{O}(1)$ using running sums of path lengths, as Figure 15 shows. After calculating the weighted sum of path lengths $p_{i,i+1}$ over the cut $(i, i+1)$, the gain can be obtained by replacing these paths (adds $p_{i,i+1}$ to the gain) with paths in the opposite direction of the cycle (subtracts $r_{\text{before}} \cdot r_{\text{after}} \cdot \ell - p_{i,i+1}$ from the gain). A factor of 2 is needed to account for the fact that each path is used in both directions.

Using this algorithm, the best exchange can be found in $\mathcal{O}(n^3)$ time, which is the same time complexity as the AMLS, but the expected time is in $\mathcal{O}(n^2 \cdot \log n)$. Since both AMLS and CYCopt pick the best pair of edges for the exchange, both lead to the same local optima. However, CYCopt is faster than AMLS, as the experiments in Section 3.3.4 will show.

RANDOM CYCLE SEARCH   To improve the complexity of CYCopt, we propose a randomized version. In CYCrand, not all $\mathcal{O}(n^2)$ cycles are checked. Instead, for every node we pick a random second node   *CYCrand*

$\textsc{ChangeRoot}(m)$ $\qquad\qquad\qquad$ *updates all $s_T(i)$, $\mathcal{O}(n)$*

$\ell \leftarrow d_{m,1}$
for $i = 1 \ldots m - 1$ **do**
$\quad \lfloor \; \ell \leftarrow \ell + d_{i,i+1}$ $\qquad\qquad$ *calculate the length of the cycle*

$c_1 \leftarrow s_T(1)$
for $i = 2 \ldots m$ **do**
$\quad \lfloor \; c_i \leftarrow s_T(i) - s_T(i-1)$ $\qquad$ *calculate the number of nodes behind i*

$r_{\text{after}} \leftarrow n$ $\qquad\qquad\qquad$ *stores the sum of demands after the cut*
$\ell_{\text{after}} \leftarrow 0$ $\qquad$ *stores the sum of half-weighted path lengths after the cut*
for $i = 1 \ldots m - 1$ **do**
$\quad | \quad r_{\text{after}} \leftarrow r_{\text{after}} - c_i$
$\quad \lfloor \quad \ell_{\text{after}} \leftarrow \ell_{\text{after}} + r_{\text{after}} \cdot d_{i,i+1}$

$r_{\text{before}} \leftarrow 0$ $\qquad\qquad\qquad$ *stores the sum of demands before the cut*
$\ell_{\text{before}} \leftarrow 0$ $\qquad$ *stores the sum of half-weighted path lengths before the cut*
for $i = 1 \ldots m - 1$ **do**
$\quad | \quad r_{\text{before}} \leftarrow r_{\text{before}} + c_i$
$\quad | \quad r_{\text{after}} \leftarrow n - r_{\text{before}}$
$\quad | \quad \ell_{\text{before}} \leftarrow \ell_{\text{before}} + r_{\text{before}} \cdot d_{i,i+1}$
$\quad | \quad \ell_{\text{after}} \leftarrow \ell_{\text{after}} - r_{\text{after}} \cdot d_{i,i+1}$
$\quad | \qquad\qquad\qquad$ *calculate the sum of weighted path lengths:*
$\quad | \quad p_{i,i+1} \leftarrow \ell_{\text{before}} \cdot r_{\text{after}} + r_{\text{before}} \cdot \ell_{\text{after}}$
$\quad | \qquad\qquad$ *calculate the gain $g_{i,i+1}$ of cutting edge $(i, i+1)$:*
$\quad \lfloor \quad g_{i,i+1} \leftarrow 2 \cdot (2 \cdot p_{i,i+1} - r_{\text{before}} \cdot r_{\text{after}} \cdot \ell)$

Figure 15: Calculating the gain of removing an edge from the cycle. The cycle is induced by edge $e = (1, m)$ and consists of the nodes $1 \ldots m$, in this order. The gain for removing edge $(i, i+1)$ is stored in $g_{i,i+1}$. All operations can be performed by iterating from node 1 to m using the parent pointers. Changing the root to m and calculating the sizes of the subtrees $s_T(i)$ is supposed to be done in an outer loop.

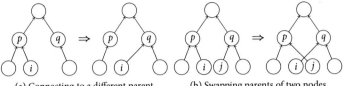

(a) Connecting to a different parent   (b) Swapping parents of two nodes

Figure 16: Two types of subtree moves

and check whether connecting these two nodes helps improving the cost. Thus, only $\mathcal{O}(n)$ cycles are checked, leading to the time complexity of $\mathcal{O}(n^2)$ for searching the neighbourhood of CYCrand, or $\mathcal{O}(n \cdot \log n)$ in the average case. This speed-up comes at a price: As the random version does not cover the whole neighbourhood of the full version CYCopt, it produces weaker results.

SUBTREE SEARCH   The subtree searches are based on an attempt to further reduce the complexity of the cycle search. Instead of considering all edges from the cycle, the subtree searches only consider the edge adjacent to the newly inserted edge. Thus, a node is disconnected from its parent and reconnected to another part of the tree, taking its subtree along.   *subtree search*

The subtree searches use two types of moves shown in Figure 16. In move (a), a subtree is moved to a different part of the tree. We improved upon an early formulation in [112] by allowing node i to connect to a node q in its own subtree. To avoid creating cycles, we have all parent pointers point in the direction of q. In move (b), two subtrees are swapped. The gain g of these moves can be calculated easily when using the objective formulation from (3.2). For both moves, only those $C_T(v)$ have to be recalculated that do change. Affected are all nodes for which the number of children or the parent edge changes, i. e. all nodes on the path P in the tree T from i to the new parent q (or from i to j for move (b)):

$$g = C(T) - C(T') = \sum_{v \in P} \left(C_T(v) - C_{T'}(v) \right) \qquad (3.11)$$

The number of updates is expected to be in $\mathcal{O}(\log n)$, but in the worst case, if the tree degenerates to a single path, it is linear in n. If we assume an expected path length of $\mathcal{O}(\log n)$, the time to search the complete neighbourhood becomes $\mathcal{O}(n^2 \log n)$, which is the same as the cycle search. Since cycle search already covers move (a), the subtree search seems weaker. But cycle search does not cover move (b), which can be seen as two moves of type (a) applied consecutively. This difference can give the subtree search an advantage. We will call this local search ST2opt.

ST2opt

REDUCED SUBTREE SEARCHES   As the complexity of the full subtree search is still $\mathcal{O}(n^2 \log n)$, a local search based on fully searching this neighbourhood does not really scale well with the problem size. Therefore, we propose two variants of the local search.

The first variant considers only the nodes adjacent to the current parent p as candidates for the new parent q in move (a). With this restriction, the path P in (3.11) consists of only two edges $\{i, p\}$ and $\{p, q\}$. A move of type (b) is performed only if $s_T(i) = s_T(j)$, which means that no update of $C_T(v)$ is necessary along the path from p to q. The gain of this move can be derived solely from the distances d_{ip}, d_{iq}, d_{jp}, and d_{jq}, and the subtree size $s_T(i)$. Furthermore, not all pairs (i, j) are considered here. Instead, in each iteration of the local search, for each node i a node j is selected randomly.

STconst

With these restrictions, the time complexity for searching the neighbourhood reduces to $\mathcal{O}(n \cdot \text{degree})$. If we assume that the node degree in the tree is bounded by a constant, this reduces to $\mathcal{O}(n)$. With this assumption, the computation time for checking one subtree to be moved is constant. We therefore call this local search STconst.

Since a local search based only on local changes and a few non-local changes that do not change s_T will most probably get stuck too early in local optima, we consider another neighbourhood. In addition to STconst we also check for non-local moves of type (a) by randomly selecting a new parent q for each node i. This way, the time required for searching the neighbourhood remains sub-quadratic (expected: $\mathcal{O}(n \cdot \text{degree} + n \log n)$) but global moves are also considered to a certain degree. We call this neighbourhood STrand.

STrand

3.3.3   Problem Instances

In order to evaluate the ELS, we use real world problem instances. A vast source of delay measurements in the Internet is PlanetLab [26], a platform connecting computers around the world to form a global research network for the development of new network services. Each participating university or research institute contributes at least two computers. As of 2010, PlanetLab consists of more than 1000 computers at more than 400 sites. In [11], the interdomain connectivity *PlanetLab* of PlanetLab is analysed. Based on Round Trip Time (RTT) measurements taken in November 2003, the authors show that triangle inequality violations are frequent even in such a small and controlled environment as the PlanetLab. The RTT measurements are accredited to Jeremy Stribling, who provided RTT measurements between PlanetLab hosts taken every 15 minutes from February 2003 to December 2005 [168]. Each link was measured using ten ICMP ping messages, and values for the minimal, average, and maximal RTT are given. For our experiments, we used the first measurement for each month of the year 2005 (denoted by 01-2005, 02-2005, ..., 12-2005), taking the minimal RTTs rounded to milliseconds [ms] as communication costs for the links. The measurements are flawed, not all hosts in PlanetLab were reachable all the time, in some cases measurements for single edges are missing for no obvious reasons. This is not necessarily a fault in the *missing links* measurement method, as in some cases these nodes really cannot contact each other directly, a behaviour that can be found quite often in PlanetLab. The authors of [15] called this phenomenon *non-transitive connectivity*. After eliminating hosts that did not report any measurements or that reported only a few measurements, the reduced distance matrices consist of 70 to 419 nodes.

Table 2 gives an overview of the PlanetLab instances. The percentage of missing links varies between 3.2 % and 7.8 %. Triangle inequality violations are very frequent, between 64 % and 86 % of all links do not *triangle* represent the shortest path between the corresponding nodes. Column *inequality* APSP shows the average communication cost (i. e. the average RTT) *violations* when only shortest paths are used for communication. This is the best communication cost achievable by any topology and therefore a valid lower bound for the MRCST. Column *Direct* gives the average communication cost when only direct links are used. Because of the triangle inequality violations this value is higher. Comparing this column to the APSP shows the degree of triangle inequality violations in the networks. Using shortest paths instead of direct links helps reducing the communication cost by at least 12 % (04-2005) and up to 69 % (12-2005) in these networks.

Although average RTTs are an intuitive measure for networks, we will use the total communication cost, i. e. the sum of all-pairs delays, for comparison to the MRCST objective defined in (3.1). Table 3 shows

Instance	Size	Missing [%]	TIV [%]	avg RTT [ms]	
				APSP	Direct
01-2005	127	3.2	79.8	152.9	249.8
02-2005	321	6.0	83.5	154.5	187.0
03-2005	324	7.8	84.3	164.0	202.7
04-2005	70	3.7	64.2	137.3	156.0
05-2005	374	6.3	86.0	124.2	192.7
06-2005	365	7.1	85.3	137.5	202.4
07-2005	380	3.8	85.1	172.7	256.0
08-2005	402	3.8	84.0	171.5	232.1
09-2005	419	5.9	85.5	132.4	222.1
10-2005	414	5.8	83.8	165.8	249.2
11-2005	407	5.5	83.2	143.4	199.4
12-2005	414	6.0	83.4	119.0	384.0

Table 2: Overview of the PlanetLab instances. For each instance, the network size, the percentage of missing links (Missing), the percentage of links violating the triangle inequality (TIV), and the average Round Trip Time (RTT) for communication using shortest paths (APSP) or direct connections (Direct) are shown.

these costs. Since the networks contain missing links, the value in column *Direct* is an underestimation of the real value. Also shown in this table are the costs for the SPT rooted at the median, which gives an upper bound for the MRCST. As shown in Section 3.2, the SPT rooted at the median is also a 2-approximation to the MRCST. With the exception of only three instances, it already performs better than the direct connection topologies. The last column in Table 3 gives the best known solution for the MRCST. In only one of the considered instances this gives a slightly worse cost than using direct connections (04-2005).

3.3.4   *Experiments*

setup

We performed several experiments to evaluate the effectiveness of the different local search variants as well as the ELS. All CPU times reported in this section refer to a Xeon E5420 2.5 GHz, running Linux. The

Instance	Size	APSP [ms]	Direct [ms]	SPT [ms]	Best [ms]
01-2005	127	2 447 260	3 868 930	3 025 120	2 743 556
02-2005	321	15 868 464	18 045 792	19 607 936	17 567 152
03-2005	324	17 164 734	19 551 610	21 117 374	19 288 320
04-2005	70	663 016	725 532	787 856	751 404
05-2005	374	17 324 082	25 198 386	19 949 036	19 175 890
06-2005	365	18 262 984	24 982 820	22 217 312	20 312 884
07-2005	380	24 867 734	35 449 904	29 288 762	27 731 218
08-2005	402	27 640 136	36 000 372	32 209 226	30 540 984
09-2005	419	23 195 568	36 620 526	27 204 468	25 712 960
10-2005	414	28 348 840	40 139 262	33 639 378	31 195 732
11-2005	407	23 694 130	31 144 230	29 322 480	26 797 284
12-2005	414	20 349 436	61 751 346	25 985 050	22 723 454

Table 3: Communication costs for the PlanetLab instances using either shortest paths for all connections (APSP), direct connections (Direct), the SPT rooted at the median node (SPT), or the best known solution for the MRCST (Best).

algorithms were implemented in C++. All results are averaged over 30 runs.

In a first set of experiments, we varied the number of generations κ, the number of offspring λ, and the local search used in the ELS. The results for these experiments are shown in Tables 4 and 5. These results are averaged over all 12 PlanetLab instances and all 30 runs per instance. To allow for quick comparison, we print the same number of digits for the values in each column. In the following paragraphs, we will discuss the different parts of this table in more detail.

Iterated Local Search (ILS)

When setting the number of offspring to $\lambda = 1$, the ELS behaves like an ILS. Considering only the subtree searches, it can be seen that with increasing neighbourhood size $|N_{\text{STconst}}| < |N_{\text{STrand}}| < |N_{\text{ST2opt}}|$ both runtime and solution quality increase. The weakest local search, ST-const, performs very poorly with a high average gap of more than 12 % from the best solution even after $\kappa = 1000$ iterations. The slight modi-

LS	λ	$\kappa = 1$			$\kappa = 10$		
		Time [s]	Gap [%]	Best	Time [s]	Gap [%]	Best
STconst	1	0.01	33.08507	0	0.2	30.21057	0
	10	0.18	27.64882	0	1.8	24.99402	0
	100	1.68	23.39780	0	18.3	20.72602	0
STrand	1	0.07	12.09002	0	0.3	5.89054	0
	10	0.33	7.01245	0	3.0	3.62344	0
	100	3.14	4.50096	0	29.9	2.70591	0
CYCrand	1	< 0.01	10.62337	0	0.1	5.60633	0
	10	0.08	6.34291	0	0.7	3.71235	0
	100	0.84	4.44986	0	6.7	2.76524	0
ST2opt	1	0.34	0.97014	19	1.5	0.32805	107
	10	1.93	0.18579	60	14.0	0.01813	312
	100	17.78	0.00894	199	133.7	0.00007	358
CYCopt	1	0.23	0.52725	55	1.1	0.11449	187
	10	1.41	0.04615	162	10.3	0.00066	351
	100	12.93	0.00050	307	99.2	0.00000	360
AMLS	1	128.45	0.51195	56	425.6	0.16368	149
	10	739.17	0.08945	108	4191.5	0.00312	291
	100	6731.06	0.00325	213	39296.8	0.00007	356

Table 4: Results for the MRCST ELS, Part 1. For the various local searches, numbers of iterations κ, and numbers of offspring λ, the CPU times and gaps to the best known solutions are shown. These values are averaged over all 12 instances and 30 runs each, totalling 360 runs for each parameter set. The columns *Best* give the number of runs that found the best known solution.

LS	λ	κ = 100			κ = 1000		
		Time [s]	Gap [%]	Best	Time [s]	Gap [%]	Best
STconst	1	0.5	21.59562	0	2.1	12.81572	1
	10	4.8	12.19508	0	18.9	3.01960	14
	100	52.8	3.86677	7	184.5	0.77779	32
STrand	1	0.8	1.75661	2	3.1	0.76131	15
	10	8.2	0.45173	26	30.3	0.25334	56
	100	86.9	0.19455	52	306.4	0.15395	119
CYCrand	1	0.2	1.63428	0	0.4	0.83222	5
	10	1.3	0.61458	0	3.6	0.36241	30
	100	13.9	0.24711	30	35.1	0.16320	38
ST2opt	1	4.4	0.25586	159	29.3	0.17630	213
	10	43.1	0.01499	328	290.4	0.01135	336
	100	420.4	0.00007	359	2914.7	0.00007	359
CYCopt	1	3.6	0.08862	244	24.9	0.05947	277
	10	34.7	0.00063	353	247.8	0.00062	354
	100	341.1	0.00000	360	2477.1	0.00000	360
AMLS	1	542.0	0.15439	165	1154.5	0.14654	179
	10	5456.0	0.00215	304	11922.3	0.00196	320
	100	50889.9	0.00004	358	114978.7	0.00004	358

Table 5: Results for the MRCST ELS, Part 2. For the various local searches, numbers of iterations κ, and numbers of offspring λ, the CPU times and
gaps to the best known solutions are shown. These values are averaged
over all 12 instances and 30 runs each, totalling 360 runs for each parameter set. The columns *Best* give the number of runs that found the
best known solution.

fication of STrand compared to STconst has a tremendous effect on the solution quality. STrand reduces the initial gap by a factor of 3, and the final gap at $\kappa = 1000$ iterations by a factor of 16, using only about 50 % more CPU time.

The random variants for cycle search and subtree search are quite comparable. STrand takes longer, but produces slightly better results than CYCrand. Since cycle search is stronger than subtree search, this difference is due to the exchange moves in the subtree search. Both ran-

dom searches greatly reduce the computation times of their full counterparts, and still find good solutions with less than 1 % deviation from the best solutions after $\kappa = 1000$ iterations.

For the stronger local searches (ST2opt, CYCopt, and AMLS), this solution quality is already reached after the first iteration ($\kappa = 1$), and it is further improved during the following iterations. With the full variant, the subtree search is weaker than the cycle search, while both show roughly the same computation times. Interestingly, CYCopt is

not only faster than AMLS, but also finds better solutions with about half the gap to the best known solutions, thus making it the best choice for finding best solutions.

During ILS mode, no local search is able to find all best known solu-

tions in all runs even with $\kappa = 1000$ iterations. CYCopt finds the best known solution in about $3/4$ of the runs, ST2opt finds them in about $7/12$ of the runs, and AMLS finds them in almost $1/2$ of the runs, with a slightly better gap than ST2opt.

Random Restart

Using only $\kappa = 1$ iteration, the ELS produces the same results as random restart with $\lambda + 1$ iterations. For the weaker local searches (STconst, STrand, and CYCrand), random restart takes longer and produces worse results than the ILS (comparing $(\lambda, \kappa) = (100, 1)$ with $(1, 100)$). For the stronger local searches (ST2opt, CYCopt, and AMLS), the opposite can be observed: Random restart still takes longer, but produces better results than ILS. However, random restart still fails to find

the best known solutions in more than one third of the runs for ST2opt and AMLS, and about one sixth of the runs for CYCopt.

Evolutionary Local Search (ELS)

When using the full ELS with $\kappa > 1$ and $\lambda > 1$, we can see that increasing the number of offspring λ significantly increases the average solution quality. The results clearly indicate that the ELS using the full versions of the local searches performs best but the random variants can also achieve very good results in less time. Even the weakest local *random vs. full* search, STconst, can be used to find solutions with less than 1 % devi- *search* ation from the best known solution. Both randomized local searches and especially STconst also scale better with the problem size.

All considered local searches in their full versions with $\kappa = 100$ generations and $\lambda = 100$ offspring per generation find the best solutions very frequently. The best local search, CYCopt, found all best solutions *best solutions* in all 360 runs in at most $\kappa = 5$ iterations, taking an average of 18 s and *found* at most 146 s. AMLS failed to find the best known solutions in two runs, and the weaker local search ST2opt failed in only one run but produced a higher gap.

A statistical analysis with 95 % confidence shows no significant difference between the quality of these local searches with $\kappa \geq 10$ and *statistical* $\lambda = 100$, where almost all runs found the best solutions. In all other set- *significance* tings, ST2opt was significantly worse than CYCopt and AMLS, except for $\kappa = 1000$ where the difference to AMLS is not significant. CYCopt was also significantly better than AMLS with $\kappa = 1$ and $\lambda \geq 10$, or with $\kappa \geq 100$ and $\lambda = 1$.

The disadvantage of these full local searches is their high complexity $(\mathcal{O}(n^3))$. With $\kappa = 1000$ iterations and $\lambda = 100$ offspring, their running *computation* times can already be measured in hours. When considering larger net- *time vs.* works, or when aiming for good solutions within seconds, the random- *solution quality* ized local searches CYCrand or STrand (both $\mathcal{O}(n^2)$) should be used in an ILS with $\kappa = 1000$ iterations. Although they tend to converge to lower-quality solutions, they already provide reasonably good results when given less time than one iteration of their full counterparts. The weakest local search STconst should only be used to improve approximations for very large instances, as it results in extraordinarily high gaps when starting from random solutions, but it shows the best complexity $(\mathcal{O}(n \cdot \text{degree}))$.

Fitness Distance Correlation (FDC)

To evaluate the potential of the local searches, we analysed the properties of the local optima. For each local search, we searched for 1000

LS	Excess [%]			FDC			Distance		
	min	avg	max	min	avg	max	min	avg	max
STconst	14.3	39.9	69.4	0.05	0.15	0.43	35	197	248
STrand	5.5	14.2	29.2	0.10	0.43	0.72	28	142	180
CYCrand	6.8	13.0	25.1	0.14	0.53	0.77	42	194	240
ST2opt	0.9	1.8	2.4	0.40	0.63	0.87	17	53	75
CYCopt	0.4	1.1	1.8	0.53	0.72	0.92	12	45	74
AMLS	0.4	1.0	1.6	0.38	0.73	0.93	13	46	69

Table 6: Properties of the local optima. For each local search neighbourhood, the minimal, average, and maximal Fitness Distance Correlation (FDC) coefficient for the considered PlanetLab instances is
shown.

local optima, starting from 1000 random solutions, and measured the
percentage excess over the best solution and the distance in the solution space to the best solutions (i. e. the number of edges not found in
any of the best solutions). From these values we calculated the Fitness
Distance Correlation (FDC) coefficient of the local optima for each
problem instance. The results are displayed in Table 6, giving the minimal, maximal, and average values for all 12 PlanetLab instances. For
detailed values on each problem instance, please refer to Table 30 in the
appendix.

When comparing the average excess of the local searches, the same

fitness conclusions can be drawn as from the ELS results. The simpler local
searches (STconst, STrand, CYCrand) lead to local optima with higher
gaps to the best known solutions. The more complex local searches (ST-
2opt, CYCopt, AMLS) give smaller gaps. CYCopt and AMLS produce
roughly the same results with about half the gap of ST2opt, but no local
search can find the global optimum in all runs without an ELS.

The distance to the best solutions in the solution space indicates how
many changes have to be applied to the local optima in order to reach a

distance best solution. It is bounded by the number of edges $n-1$ in the solution
tree. In the considered PlanetLab instances, the size varies between 69
and 418, with an average of 333.75. The values for the average distance
for all local searches show that the local optima are still far from the best
solutions in the solution space. On average, between every sixth and

every seventh edge has to be changed to reach the best known solutions when starting from the local optima produced by ST2opt, CYCopt or AMLS. For the random variants, on average about half of the edges have to be changed. An interesting side note is that the local optima of ST-rand are closer to the best known solutions. Even though this does not result in better gaps than CYCrand, it still shows the influence of move (b) from Figure 16, which is not covered during a CYCrand search.

Column FDC gives the correlation between distance and solution quality. If the correlation coefficient is positive, shorter distances correlate with better solutions. For all considered local searches and problem instances, the FDC coefficient is positive, indicating that an EA can exploit the structure of the search space.

correlation

For the weaker local searches, the average FDC coefficient is very low, indicating that recombination may not be very helpful in these landscapes, as respectful recombination is known to be effective on correlated landscapes [109].

The FDC coefficients for the stronger local searches is higher. These values as well as the FDC scatter plots (e. g. Figure 17) indicate that the fitness landscapes appear to be correlated, and a recombination might be effective. However, as the ELS already finds the best known solutions in almost all runs using the stronger local searches, recombination seems unnecessary.

The scatter plots in Figure 17 are typical for about half of the PlanetLab instances. In the remaining instances, the scatter plots for the random variants STrand and CYCrand are not split, but form one connected cloud. The plots for AMLS and CYCopt show that these strong local searches produce fewer different local optima than ST2opt. Comparing the plots for AMLS and CYCopt also shows that these local searches lead to different local optima, although both searches should be exchangeable. However, when they find two moves during their search that yield the same gain, AMLS and CYCopt will select differently, because they check the moves in a different order.

Best known solutions

The best known solutions for the PlanetLab instances have been found quite regularly by the ELS with ST2opt, CYCopt, and AMLS using only moderate parameter settings. Since even higher parameter settings with the strongest local searches did not find any improvements, we

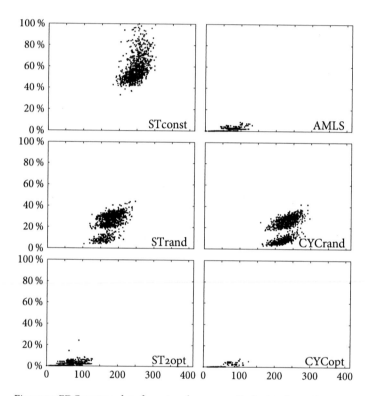

Figure 17: FDC scatter plots for network 12-2005. Each plot shows the excess and the number of different edges for 1000 local optima of one local search. The subtree searches on the left are sorted by their strength: STconst, STrand, ST2opt. The cycle searches CYCrand and CYCopt are placed next to the corresponding subtree variant. AMLS on the top right should give the same local optima as CYCopt on the bottom right.

believe these best known solutions to be optimal. The best known solutions have already been shown in Table 3 on page 51.

In this table, the APSP costs are lower bounds, and the SPT costs are upper bounds to the MRCST. The costs for directly connecting all nodes (Direct) could also serve as a lower bound, if there were no triangle inequality violations in the networks. However, for real world networks as the considered PlanetLab instances, the triangle inequality is frequently violated, and the best known solutions for the MRCST are often better than the direct connection. Only in the smallest instance 04-2005, the best solution found by the ELS gives slightly worse results.

direct connection is not the best solution

The APSP measures the overall communication cost if every communication is routed over the shortest path instead of a confining all communication in the network to a single shared tree as in the MRCST. The gap between the best solutions for the MRCST and this simple APSP lower bound is relatively small: between 10.0 % and 13.3 %. This indicates that the advantage of using only one tree in a P2P network does not need to come at a high price. Instead of determining all shortest paths and maintaining a high number of links ($\mathcal{O}(n^2)$), a single tree as the MRCST only uses $n - 1$ links in the network.

APSP

Comparing the best known solutions to the upper bound given by the SPT rooted at the median node shows that the MRCST can improve these simple approximations by between 4 % and 14 %. Although the SPT upper bound can be found easily and the MRCST is an $\mathcal{NP}$-hard problem, using the ELS one can find tree topologies with less than 1 % deviation from these best known solutions in less than a second.

what can be gained

3.3.5 Discussion

In this section, we presented a highly effective Evolutionary Algorithm (EA) incorporating local search for the Minimum Routing Cost Spanning Tree (MRCST). This Evolutionary Local Search (ELS) was tested on test instances based on real world Internet measurements. In these networks, the MRCST topologies come close to a lower bound defined by using shortest paths for all connections, and they improve upper bounds defined by the best Shortest Path Tree (SPT).

ELS

We investigated three types of local searches, a simple subtree search, a cycle based search, and an adapted OCST search, in three variants, full search, randomized search, and reduced local search, in the evolutionary framework using a self-adapting mutation operator. The sub-

local searches

tree searches have already been published in [112], and the full CYCopt has already been published in [184]. The results show that the ELS when used with the full CYCopt local search finds the best known or optimal solutions in short time. The results also show that using randomized local searches with lower time complexity yields very good solution qualities, an excess of only 0.5 % can be reached in a couple of seconds for the considered network sizes.

The results of the ELS have shown that heuristics can be efficiently used to solve the MRCST. However, they cannot be used to prove the optimality of the created solutions. There are two possible approaches to achieve this: Exact algorithms based on Mixed Integer Programming (MIP) formulations for the MRCST could be used, such as the algorithms proposed by Fischetti *et al.* in [54]. However, as their running times increase exponentially with the problem size – the MRCST is $\mathcal{NP}$-hard – these algorithms take an exceedingly long time even for the smallest PlanetLab instance considered in this chapter. A second approach to show the quality of the solutions found by the ELS is to use better lower bounds. The APSP lower bound does not honour the tree structure of the MRCST and produces very low bounds. An attempt to relax the tree constraint ($n - 1$ edges) in a simple MIP formulation for the MRCST resulted in either very similar lower bounds as the APSP ($\approx n^2$ edge) or in unacceptably long running times ($\approx n$ edges) or both (in between). When using the PTAS proposed by Wu *et al.*, the cost C of the resulting $(1 + \epsilon)$-approximation could be used to calculate a lower bound $C \cdot (1 + \epsilon)^{-1}$, but the running times are again unacceptably high at the accuracy level needed for the solution quality provided by the ELS.

no proof for optimality

3.4 DISTRIBUTED ALGORITHM

In a P2P system that uses a spanning tree topology as the overlay network, this overlay topology could be optimized by gathering all necessary data in one single node, having this node execute a global view algorithm for the MRCST as presented in the previous section, and sending the results back to all other nodes. However, this approach is futile for two reasons: Firstly, the underlying network's properties and even the P2P system may change during the time it takes to gather all data, calculate a good solution, and broadcast the results. Secondly, and more importantly, gathering all data is a very expensive task, since the RTT between each node pair has to be measured.

TreeOpt, the algorithm presented in this section, is a fully distributed algorithm, where each node in a self-organizing P2P system acts on its local view only. We will show that this approach reduces the necessary degree of cooperation, and leads to reasonably good results. With this approach, the P2P system will be used as a self-optimizing system. The distributed algorithm works by evolving Shortest Path Trees (SPTs) in which the root is allowed to move.

This section is organized as follows. Related approaches for finding optimal tree topologies in P2P systems are discussed in Section 3.4.1. Our distributed algorithm for tree optimization is presented in Section 3.4.3. In Section 3.4.4, results from experiments based on measured Internet data are shown. Section 3.4.5 concludes this section on the distributed optimization of the MRCST.

3.4.1 Related Work

Modelling P2P systems as self-organizing networks helps to understand their ability of dealing with concurrent joining and leaving of peers without relying on a central component. Thus, the peers in a P2P system are autonomous interacting entities forming a complex dynamic system. The P2P system can also be used as a self-optimizing system, if appropriate distributed algorithms are used. One optimization goal in P2P systems is the communication cost. Established P2P protocols like CAN [147], Chord [167], Pastry [151] or Gnutella [150] are self-organizing: They already deal with the dynamics of joining and leaving of peers. However, they are not self-optimizing, as they do not consider the end-to-end message delays. Systems like Gnutella or Pastry already prefer links with low RTTs, which reduces the load on the underlying network, but they do not explicitly optimize overall communication costs. *P2P systems*

Only few research has been conducted on the optimization of network topologies, in particular P2P overlay topologies. In [164], Ahmed Sobeih *et al.* compare the performance of tree and ring topologies in overlay networks. Although they formulate both topologies as $\mathcal{NP}$-hard optimization problems (a Steiner tree problem and a variant of the Traveling Salesman Problem), they use very simple global view heuristics (MST and nearest-neighbour) to find approximate solutions. Based on these solutions, they conclude that ring topologies may not be suitable for delay-sensitive applications, as they induce a path stretch of up to an order of magnitude larger than a tree topology. *ring vs. tree (decentralized) EA*

Katharina A. Lehmann and Michael Kaufmann propose a very general EA framework for self-organized networks in [97]. The framework can be seen as an adaptation from a $(1+1)$-EA, as at any time only one solution is kept. Mutation is done in steps. In each time step, only one node is active. This randomly chosen node can insert or remove edges based on its local knowledge. It remains unclear how their algorithms can work in a fully decentralized distributed system. The network problems considered in [97] are simple optimization problems, in that local changes have only local effects. This is not the case for the MRCST, where a simple edge swap affects many different communication routes and thus can lead to an unexpected increase in the total communication cost.

application-layer multicast

In [16], Eli Brosh and Yuval Shavitt present a centralized approximation algorithm as well as a centralized heuristic for application-layer multicast trees. The paper concentrates on the $\mathcal{NP}$-hard optimization problem to find a minimum delay multicast tree. In addition to the transmission delays on the edges of the network, a non-negative processing delay is assigned to each node. This problem is related to our TreeOpt approach as the authors are minimizing the communication delay and at the same time reduce the internal node degree. However, their objective is different, since they only consider the multicast tree from a single source, whereas the MRCST minimizes all communication in the network.

Mesh-Tree

The MeshTree approach presented by Su-Wei Tan *et al.* in [171] is a decentralized approach to find degree-bounded MSTs to be used as multicast trees. In a first step, all nodes in the network build a degree-bounded mesh that contains a low cost backbone tree. The actual delivery trees used as multicast trees differ from this backbone tree, but also use only edges from the mesh. It is unclear how effective the optimization is in terms of approaching the optimum or a lower bound.

AMLS-like approach

In all application-layer multicast approaches, a different tree is built for each source node. In the MRCST, we will use a single tree to route all communication. One approach to create a distributed algorithm for the MRCST is introduced by Linda Pagli and Giuseppe Prencipe in [136]. They present a fully distributed algorithm to find the best replacement edge for an edge of the tree. Their algorithm uses $\mathcal{O}(n \cdot m)$ messages to find these replacement edges for all edges of the tree. Here, m denotes the number of edges in the network. In P2P networks, the number of edges is $m = n^2 - n$. In a way, this distributed algorithm can be seen as a single step of the AMLS. However, the step is not taken, the replacement edges are just stored. The actual swap is done when an edge of the tree breaks down due to a link failure. In order to build a distributed algorithm for the MRCST, the algorithm would have to detect termination, find the best swap out of all these replacement edges, swap these edges, and repeat this process as long as the best swap improves the overall communication cost. Since all nodes of the network have to cooperate in each step, this approach is not suitable for larger networks. A simple improvement would be to use a form of cycle search, where a new edge is inserted and only the nodes along the induced cycle need to cooperate. However, the level of cooperation would still be high with $\mathcal{O}(\log n)$ in the average case.

3.4.2   From Global Optimization to Distributed Algorithms

In Section 3.3.2, we presented powerful local searches based on global view optimization. All these local searches could be implemented in a distributed self-organizing way. However, calculating the gain of a move requires cooperation of many nodes in the network. For AMLS, all nodes need to cooperate to calculate the w_i and h_i values. For the cycle searches, all nodes in a cycle need to cooperate. For the subtree search variants ST2opt and STrand, when a node i wants to connect to a new parent q, all nodes on the path from i to q need to cooperate. This path length and the cycle length is only bounded by the number of nodes in the network. Limiting the allowed path or cycle length and hence the number of cooperating nodes may reduce the effectiveness of the local search. Although the weakest search STconst does not need many cooperating nodes, it also produces very weak solutions, so it is not a good choice for a distributed algorithm either.

level of cooperation

One important aspect of self-organizing networks is their ability to create a global structure based on local rules. When developing a distributed algorithm for creating a self-organizing network, it is therefore a valid goal to reduce the necessary amount of cooperation. We investigated the importance of cooperation in the local search for the MRCST by restricting the maximum path length of the moves and used a path length restricted ST2opt to find 1000 local optima for each PlanetLab network. The results of these preliminary experiments are shown in Figure 18, averaged over all instances. The experiments show that for the PlanetLab networks this path length already has to be as high as 10 in order to arrive at near optimum solutions. These results motivated to find alternative distributed algorithms for the MRCST.

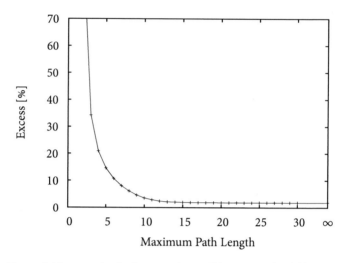

Figure 18: Necessary level of cooperation in ST2opt. For the different maximum path lengths, the excess over the best known solution is shown. The values are averaged over all PlanetLab instances.

3.4.3   TreeOpt: A Distributed Algorithm Based on Shortest Path Trees

The basic idea behind the TreeOpt algorithm is as follows: Rather than performing a local search on the MRCST directly, we perform a local *alternative* search to find SPTs. SPTs have been successfully used as approximations to the MRCST and can be found more easily in a distributed system. Compared to the local searches for the MRCST, TreeOpt requires a minimum amount of cooperation of the peers. Each peer tries to improve its situation in a greedy, selfish fashion.

TreeOpt also includes mechanisms to find the median node in the network and strives to use this node as the root node of the constructed SPT.

Since tree topologies suffer hugely when a link is broken, TreeOpt introduces a network repair mechanism to reconnect the tree parts. In order to reduce the time to repair the network, TreeOpt limits the maximum number of children a node can accept (degree) to a small constant value.

This section quickly recapitulates SPTs and then describes the different parts of TreeOpt in more detail.

On Shortest Path Trees

The SPT consists of shortest paths from a given root node to all other nodes in the network. As shown in Section 3.2, the SPT rooted at the median node already gives a 2-approximation for the MRCST [190]. Because of this close relation between SPT and MRCST, minimizing *SPT and* the path from any node to the root node also helps reducing the cost *MRCST* function of the MRCST.

The SPT can be formulated as an optimization problem to find a spanning tree rooted at a given node $r \in V$ that minimizes the sum of *finding SPTs* distances from this node to all other nodes:

$$C_{\text{SPT}}(T) = \sum_{v \in V} d_T(r, v) \qquad (3.12)$$

The SPT can be solved in polynomial time using e. g. Dijkstra's algorithm [44] or the algorithm of Bellman and Ford [13, 55]. However, the resulting trees may degenerate to stars. In the context of P2P overlays, this would put a high load to the root node. We therefore propose to *load-relief* use degree-constrained SPTs, where all nodes have at most k children. In P2P systems, the degree k should be kept small since on failure of a peer, each of the k children has to find a new parent.

An SPT can also be used for application-layer multicast. The following function

$$C_{\text{multicast}}(T) = \max_{v \in V} d_T(r, v) \qquad (3.13)$$

denotes the time or cost required for a multicast from the root node r to all other nodes. The cost of a multicast from any other node is equal *multicast* to or smaller than $2 \cdot C_{\text{multicast}}(T)$.

Before explaining the SPT-based TreeOpt in detail, we use the results from the global view optimization to evaluate the quality of SPT-approximations and therefore the best possible quality of the TreeOpt *best case* approach. Table 7 shows the best solution found by the ELS presented *analysis* in Section 3.3.2, the All-Pairs-Shortest-Path (APSP) lower bound, and the Shortest Path Tree (SPT) upper bound for all of the considered PlanetLab instances. This upper bound is also a 2-approximation for the MRCST.

Instance	Size	Best MRCST	Lower Bound (APSP)	Upper Bound (SPT)
01-2005	127	2 743 556	2 447 260 (12.1 %)	3 025 120 (10.3 %)
02-2005	321	17 567 152	15 868 464 (10.7 %)	19 607 936 (11.6 %)
03-2005	324	19 288 320	17 164 734 (12.4 %)	21 117 374 (9.5 %)
04-2005	70	751 404	663 016 (13.3 %)	787 856 (4.9 %)
05-2005	374	19 175 890	17 324 082 (10.7 %)	19 949 036 (4.0 %)
06-2005	365	20 312 884	18 262 984 (11.2 %)	22 217 312 (9.4 %)
07-2005	380	27 731 218	24 867 734 (11.5 %)	29 288 762 (5.6 %)
08-2005	402	30 540 984	27 640 136 (10.5 %)	32 209 226 (5.5 %)
09-2005	419	25 712 960	23 195 568 (10.9 %)	27 204 468 (5.8 %)
10-2005	414	31 195 732	28 348 840 (10.0 %)	33 639 378 (7.8 %)
11-2005	407	26 797 284	23 694 130 (13.1 %)	29 322 480 (9.4 %)
12-2005	414	22 723 454	20 349 436 (11.7 %)	25 985 050 (14.4 %)

Table 7: Gaps to upper and lower bounds. For each PlanetLab instance, the network size, the best known solution for the MRCST, the APSP lower bound and the SPT upper bound as well as the gaps to these bounds are shown.

The table shows that there are shared trees which are very close to the APSP lower bound, all deviations are less than 14 %. This increase in cost is quite low, considering the reduced effort for maintaining a single tree instead of a set of n spanning trees as in the APSP. The table also indicates that SPTs can be used to find good approximate solutions in real world instances. The gaps to the best MRCST solutions are between 4 % and 14.4 %.

TreeOpt: General Outline

The TreeOpt algorithm is a fully decentralized algorithm. Each peer runs the same algorithm and works on local view only. The peers in the network cooperate by exchanging messages and by establishing or removing links.

epidemic
algorithm

After joining the network, each peer builds and maintains a peer list using an epidemic algorithm [53]. This epidemic part serves two purposes: Firstly, it increases robustness since it is unlikely that the overlay will be partitioned in case of link or peer failures. Secondly, the peer list maintained by the epidemic algorithm is used to establish connections to random peers that are not necessarily a direct neighbours in the tree. The peer list held by each peer does not need to contain all peers. Only a relatively small number of peers with respect to the network size have to be known in order for the epidemic part to work.

The epidemic algorithm is extended such that in every round, data for the SPT optimization is updated and an improvement step is executed in which the peer tries to improve its tree connections. A network repair mechanism helps to reconnect the tree after link failures. The general outline of the TreeOpt algorithm is shown in Figure 19.

optimization steps

Joining and Leaving

A peer joins the overlay by connecting to a peer that already participates in the network. This peer either accepts the joining peer as a child in the overlay tree or redirects it to one of its children randomly if the maximum number of children is reached. During joining each peer receives a list of other peers as usual in an epidemic algorithm. In TreeOpt, the joining peer i also receives an estimation of the distance of the parent p_T to the root r: $d_T(p_T(i), r)$.

The leaving of a peer is handled by rejoining the children of the leaving peer. As TreeOpt limits the number of children for each peer, usually only one of the children can take the former place of the leaving peer. The remaining children peers are redirected to other parts in the subtree, according to the same rules as when joining new peers.

The Tree Improvement Step

In the TreeOpt algorithm the same moves are considered as for the subtree searches shown in Figure 16. In terms of P2P overlay tree construction the first move (a) can also be seen as a peer i trying to connect to a new parent q. If peer q accepts more children, and if the distance from peer i to the root is reduced, the new link is established. Denoting the distance from peer x to the root in terms of the sum of the weights of the path $P_{x \to r}$ by $d_T(x, r)$, the gain in cost of choosing peer q as the new parent for i is

reduce distance to root

$$g(i, p, q) = d(i, p) + d_T(p, r) - d(i, q) - d_T(q, r) \qquad (3.14)$$

If this gain is greater than zero, the move is accepted.

every δ time units do:

$p \leftarrow \text{RANDOMPEER}(P)$ *Epidemic*
$\text{SENDPEERLIST}(p, P)$
$P \leftarrow \text{UPDATEPEERLIST}(\text{RECEIVEPEERLIST}(p))$

$\text{UPDATEAVERAGERTT}()$ *Tree Improvement*
if $r = i$ **then**
    $x \leftarrow \arg \min\limits_{p \in P : p_T(p) = i} \text{AVGRTT}(p)$
    if $\text{AVGRTT}(x) < \text{AVGRTT}(i)$ **then**
        $\lfloor$ $\text{MOVEROOTTO}(x)$
else
    $d_T(i, r) \leftarrow d(i, p_T(i)) + d_T(p_T(i), r)$
    $p \leftarrow \text{RANDOMPEER}(P)$
    $\lfloor$ $\text{PROBEFORMOVE}(d_T(i, r), p)$

if $\textit{PARENTLOST} \wedge \textit{RECONNECTFAILS}$ **then**
    $r \leftarrow i$ *Network Repair*
    $\lfloor$ $d_T(i, r) \leftarrow 0$
if $r = i$ **then**
    $p \leftarrow \text{RANDOMPEER}(P)$
    if $r_p < i$ **then**
        $\text{CONNECTTO}(p)$
        $\lfloor$ $r \leftarrow r_p$

Figure 19: The TreeOpt algorithm run on peer i. Three parts are shown: epidemic algorithm, tree improvement and network repair. The tree improvement step and the network repair mechanism depend on whether the peer i is a root r.

The second move (b) only occurs when peer q has already reached the maximum number of children. This peer q may still decide to accept the new child i if one of its other children j has a larger distance to q than i. Thus, it searches for the child j with maximum distance to itself ($j = \arg\max d(j, q)$), and redirects it to the parent of peer i, if $d(i, q) < d(j, q)$. This means that the distance to the root of peer i is reduced, possibly at the cost of peer j.

swapping two peers

TreeOpt Messages

All steps of TreeOpt are implemented using messages. In order to estimate its distance to the root, each peer sends a PING message to its current parent, which answers with a PONG message containing its own distance to the root. The child adds the round-trip time for this message exchange and updates its estimation.

estimate distance to root

Each peer also tries to improve its situation by looking for a better parent. It again sends a PING message to a randomly chosen peer, and measures the round-trip time for this message exchange. If choosing the new parent leads to a shorter distance to the root, the peer tries to connect to the new parent using a CONNECT message. The chosen parent may either accept the new child, using an ACCEPT message, if the maximum number of children is not reached or if the candidate peer is closer than some other child. In the latter case the child with the maximum distance is redirected to the old parent of the new child using a REDIRECT message. A REJECT message is sent if the new parent does not accept the new child.

find new parents

Hence, each peer sends two PING messages and receives two PONG messages containing the distance from the old and new parent to the root. The round-trip time of these requests is measured and the peer can decide to connect to the new parent. If it decides to change the parent, it sends a CONNECT message and receives either an ACCEPT or a REJECT message from the new parent. If the new parent decides to swap a child, it sends a REDIRECT message to its child with the maximum distance, which in turn connects to the other parent using a CONNECT message. Hence, each move can be realized with six (if the peer is rejected) to nine messages (if the peer is accepted and another peer is redirected).

number of messages

In the TreeOpt algorithm, each peer acts greedily to improve itself, trying to reduce either the distance to the root or the distance to its

selfish peers

children. No further incentive mechanisms are required for the optimization to work in a P2P environment.

Moving the Root

In the algorithm presented so far, the root of the tree is formed by the first peer joining the overlay. However, if the root is not a central peer within the network, the resulting SPT will not approximate the MRCST very well. Therefore, TreeOpt tries to move the root towards the centre of the network. This is accomplished as follows. Each peer estimates its average distance to the other peers by summing up all RTT times and *finding the median node* counting their numbers. Every m rounds, the root peer considers to pass the root role to one of its children. If there is a child with a smaller average distance estimation, this child becomes the new root. This way the root gradually moves towards the centre of the network. Both peers (old and new root) update their distance to the root estimation, which will be propagated through the network by the tree improvement part of TreeOpt.

cooperation is needed In order for this mechanism to work, peers are required to cooperate. Otherwise, the root may not move at all. In this case the overlay still evolves over time – however, to a suboptimal configuration.

Self-Repair and Network Partitioning

tree deformation Due to the limited local view of the peers a peer might unknowingly choose a new parent from its own subtree. In this case, the resulting graph is no longer a tree. A cycle is introduced and this subtree is disconnected from the remaining network. To prevent this scenario, peers may gather information about the path to the root and check it for cycles. This information may be outdated, so the cycle could remain unnoticed for a short period of time. An alternative is to tolerate such moves. TreeOpt is already able to repair the tree since a cycle leads to a permanent increase of the distance to the root for all peers within the cycle. These peers will then choose alternative parents closer to the root in subsequent moves, break the cycle and reconnect the graph.

unannounced leaving Still, the unannounced leaving of a node that is not a leaf node in the tree will result in a partitioned network. When a peer notices that the connection to its parent is lost, it should assume such a network partition and start a repair mechanism. We propose a simple mechanism to repair these kinds of network partitions which works on the data available in each peer and does not need a global view or information about the path to the root as suggested earlier.

The network repair mechanism is included in Figure 19. As a first step after recognizing the partitioning, the children of the missing peer try to reconnect to peers they have recently been connected to. In most cases, this helps to keep the tree connected. However, if too many peers fail, these orphans start to act as roots of their subtrees. This also means that their estimated distance to the root is set to $d_T(i,r) = 0$. Thus, a partitioned network will consist of a forest. In order to reconnect the trees, all root peers try to find roots of other trees. Each root r randomly selects a peer i and asks for its root peer r_i. If the root peers differ ($r \neq r_i$), peer r connects to peer i. To avoid creating cycles when the root peers of two trees try to reconnect the network at the same time, an arbitrary order has to be used, so connections are allowed in one direction only. If all peers have unique IDs, the peer with a higher ID is allowed connect to a peer with a lower root ID ($r > r_i$). The result is the merging of two trees, but it will probably not represent a good overlay tree with respect to the routing cost.

network repair mechanism

tree merging

This mechanism can take more than one round to repair the network, simply because it is a stochastic algorithm. So in some cases the network will remain partitioned for more than one round of the algorithm. TreeOpt will still optimize the unconnected trees, but the routing cost for such a partitioned network is undefined, keeping in mind that it is the sum of all communication costs between all node pairs using only the edges of the tree(s).

3.4.4 *Experiments*

We performed several experiments to study the effectiveness of SPT-based optimization in P2P overlays. In these experiments, the Tree-Opt algorithm was simulated. Each peer worked on its local view only, which was updated after every time step in accordance to the message exchange described in Section 3.4.3. All results presented in this section are averaged over 64 runs.

setup

In addition to the PlanetLab instances presented in Section 3.3.3, we also took a larger network with 2500 nodes from the Meridian project [186]. As this Meridian network is by far too large to find an optimal solution, we will revert to the best SPT as a baseline for the comparisons in this section.

problem instances

When comparing the communication cost, we will use to the average message delay d_{avg}, as this gives a more intuitive measure for network latency than the total communication cost $C(T)$ used in section Section 3.3. The average message delay is calculated as

d_{avg}

$$d_{avg} = C(T) \cdot \frac{1}{n \cdot (n-1)} \tag{3.15}$$

However, the percentage excess over lower bounds or best known solutions remains the same as when using the total communication cost.

SPT based approximation

In a first set of experiments, we focus on the case where all nodes join more or less simultaneously and the peers remain active until the end of the optimization. Each node runs the TreeOpt algorithm, trying to approximate the MRCST by using SPTs. Since these trees may degenerate to stars, we limited the number of children in the tree to a predefined degree k. In Figure 20, the influence of the maximum degree k on the performance of TreeOpt is displayed for network 12-2005. All other networks showed similar behaviour. The graph displays the average communication cost of the generated tree d_{avg} over the time. There is a significant difference in cost if the degree is varied from $k = 3$ to $k = 21$. However, the gain in cost decreases quickly. For degree $k = 5$, the performance is almost as good as for degree $k = 21$ or higher. This indicates that the MRCST can be effectively approached using SPTs with a maximum degree of $k = 5$. We will therefore use this setting in all remaining experiments.

necessary degree

moving root

Using this value of $k = 5$, we studied the influence of the moving root as described in Section 3.4.3. TreeOpt was allowed to move the root every $m = 8$ time units. SPTs with an arbitrary fixed root are not expected to give a good approximation for the MRCST, especially if the root is far from the centre of the network. In Table 8, the percentage excess above the best known solution as well as the gain G given by the quotient of a random tree and the optimized tree are compared for the case with a random static root and the case where the root is allowed to move. As we give average values over 64 runs, the values for the static root scenario do not represent the worst case, where the farthest node is chosen as the root node, nor do they represent the best case, where the median node is chosen by chance. In these experiments, we allowed

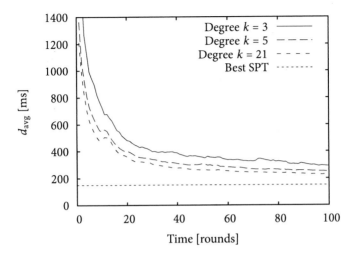

Figure 20: Influence of the maximum degree for network 12-2005. The plot shows the total communication cost as it is optimized by TreeOpt using different settings for the maximum node degree k. The SPT cost gives the best cost reachable by an SPT approximation.

TreeOpt to optimize the trees for $10 \cdot n$ rounds, where n denotes the network size.

The results demonstrate the importance of adapting the root of the tree during the TreeOpt search. For network 04-2005, moving the root reduced the gap to the best known solution from 67 % to 8 %. Moreover, the comparison to unoptimized trees indicate that the communication costs can be reduced by a factor of up to $G = 8.4$ for the considered networks using the TreeOpt algorithm.

the root needs to move

Figure 21 shows the decrease of the average communication cost over time as TreeOpt improves the tree topology for network 12-2005. The moving root scenario is clearly superior to the static case. The figure only shows the first 100 rounds of the optimization, where the effects of the adaptation to a new root can be observed. The root is moved for the first time in round 8. As the information about this root change needs to be propagated to the remaining nodes, some of these nodes will make decisions based on outdated information. During a short period of time after round 8, the communication cost of the tree in-

		Static root		Moving root	
Instance	Size	Gap (%)	G	Gap (%)	G
01-2005	127	36.09	5.3	20.97	6.2
02-2005	321	59.42	4.4	18.18	5.6
03-2005	324	68.81	4.0	15.77	6.2
04-2005	70	66.78	2.4	8.02	3.8
05-2005	374	40.08	6.4	12.85	8.4
06-2005	365	65.86	4.9	16.02	7.1
07-2005	380	33.50	5.4	14.05	6.1
08-2005	402	61.36	4.0	13.74	5.9
09-2005	419	45.45	5.9	15.97	7.8
10-2005	414	43.02	5.4	15.29	6.1
11-2005	407	50.55	4.9	17.02	6.3
12-2005	414	54.19	6.8	34.20	7.6
average:	335	52.09	5.0	16.84	6.4

Table 8: Influence of the moving root: For each PlanetLab instance the excess above the SPT and the gain G compared to unoptimized trees is shown when using TreeOpt with a static root or with a moving root.

creases. When the root is moved for the second time in round 16, this change is already too small to have a noticeable effect on the communication cost.

Scalability

In order to investigate the scalability of TreeOpt, we also tested randomly generated larger networks. In these networks, we placed n nodes in a 1000 ms × 1000 ms square and used the Euclidean distance as the RTT. These instances appear to be harder for heuristic search. The gap to the APSP lower bound is higher and the improvement of the algorithms compared to the initial random trees is lower than for the networks based on PlanetLab data.

In Figure 22, the gain G compared to unoptimized random trees is shown over the number of rounds of TreeOpt. In the experiments, the maximum degree was again set to $k = 5$ and the root adaptation was

gain

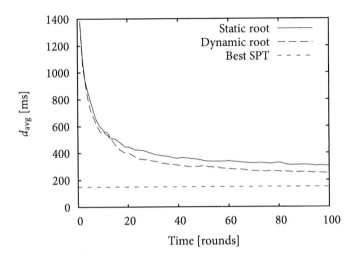

Figure 21: Influence of the moving root: The total communication cost for network 12-2005 is shown over time using either a static root or a moving root. The SPT cost gives the best cost reachable by an SPT approximation.

activated. The improvement achieved in the first rounds of TreeOpt appears to be roughly independent of the network size. The final results
indicate that the gain achieved by TreeOpt grows considerably with increasing network size, reaching values of almost 15 for a network with
one million nodes. This shows that topology optimization becomes
more important for larger networks.

Table 9 shows the gaps to the APSP lower bound for these networks
after round 100, 1000, and 10 000 of TreeOpt. The first columns show
that the initial random solution gets worse with larger networks. However, after the first 100 rounds the improvement can already be seen in
each case. The results after 1000 or 10 000 rounds show that TreeOpt
scales well with the problem size. The final values seem to be worse
than the results from real world examples (e. g. Table 8) because of the
Euclidean distances used in these randomly generated networks.

*approaching a
lower bound*

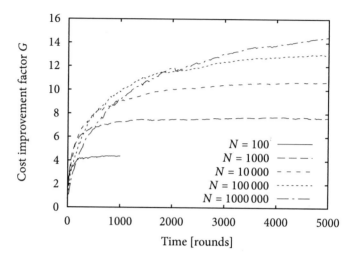

Figure 22: Improvement over random trees for large random networks. Tree-Opt was used with degree $k = 5$ and a moving root. The improvement shown is calculated by the gain compared to the initial random tree.

		Gap to lower bound after round			Lower Bound
Size	0 (Initial)	100	1000	10 000	APSP
100	577.1 %	94.1 %	55.0 %		$5.593 \cdot 10^6$
1000	1041.3 %	136.3 %	56.9 %	46.1 %	$5.244 \cdot 10^8$
10 000	1530.9 %	264.0 %	81.0 %	49.2 %	$5.230 \cdot 10^{10}$
100 000	2017.8 %	464.2 %	116.1 %	55.8 %	$5.220 \cdot 10^{12}$
1 000 000	2562.8 %	784.1 %	188.8 %	67.1 %	$5.215 \cdot 10^{14}$

Table 9: Influence of the network size on the topologies constructed by TreeOpt. For random networks of different sizes up to one million nodes, the gap to the APSP lower bound is shown for the initial random topology and after 100, 1000 and 10 000 rounds of TreeOpt using a moving root.

Churn

All previous experiments focussed on static networks where all nodes remained connected and active. However, in usual P2P environments peers are expected to join and leave at all times. To show that TreeOpt can cope with churn we conducted experiments where a large percentage of nodes suddenly leaves or joins the network. In these experiments we used the Meridian network with 2500 nodes. To make the effect of leaving nodes visible, we started with a static network for an initial phase and scheduled the joining or leaving of the nodes at some point when TreeOpt has already found a good overlay tree, but still has some room for improvements. This way we can filter out other influences like moving roots. *dynamic networks*

In the first experiment concerning churn, we were interested in long term effects of the unannounced leaving of nodes. We compare the Meridian network where 20 % of the nodes are scheduled to fail after 1000 time rounds to the same network where 20 % of the nodes are already left out and TreeOpt can optimize the routing cost from the beginning. Figure 23 shows the progress of TreeOpt for both cases. After the loss in round 1000 the tree is quickly repaired, but the routing cost increases significantly. It can be seen though that both networks, then containing the same number of nodes, converge to the same routing costs. *leaving*

In the opposite experiment we scheduled 20 % of the nodes to join after 1000 time rounds and compare this to the complete Meridian network which TreeOpt can optimize from the beginning. Figure 24 shows the results. Again a significant increase in the routing cost occurs when the new nodes connect to a random node. In this scenario, TreeOpt can optimize the tree faster than in the previous experiment. TreeOpt can now start from an already good tree as a core, instead of a number of subtrees remaining from a formerly good tree. *joining*

In the last experiment concerning churn we combine joining and leaving. Here, the network is scheduled to lose a certain amount of its nodes and at the same time receive the same number of new nodes. To be able to compare the routing cost before and after this event, we used the same RTT vector for the new nodes as for the leaving nodes. In a way, these nodes can also be regarded as the same nodes after losing all knowledge about the network. Table 10 shows the necessary repair times of TreeOpt. An exchange of only 1 % of the nodes can be repaired quickly. Even with heavy changes, the network is reconnected after at most 40 rounds. However, the resulting tree still shows a high routing *combined leaving and joining*

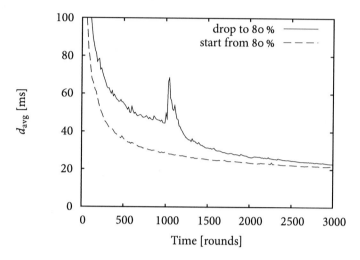

Figure 23: Effects of leaving nodes. While TreeOpt works on the Meridian network, 20 % nodes are scheduled to leave in round 1000. The average
communication cost is compared to a TreeOpt run where these 20 %
nodes are missing from the start of the optimization.

cost. Figure 25 shows the progress of TreeOpt as it re-optimizes the
tree. If 1 % of all nodes are exchanged in round 5000, the repaired tree
already shows almost the same routing cost as before. For larger losses,
the routing cost of the repaired tree is higher. However, TreeOpt can
quickly reduce the routing cost to previous levels, if up to 5 % of the
nodes leave at the same time. Losses of more than 10 % of the nodes at
the same time heavily disturb TreeOpt, as the repaired tree loses many
of the already optimized edges. With a loss of 50 %, TreeOpt almost
shows the same progress as when optimizing a totally random tree.

Noise

As TreeOpt solely bases all tree improvement decisions on RTT measurements, we conducted experiments to show how TreeOpt can cope
with noise in these measurements. Table 11 shows the results for a random Euclidean network with 10 000 nodes. These nodes were again
placed in a 1000 ms × 1000 ms; in fact it is the same network as in the
scaling experiment. A random error has been added for each meas-

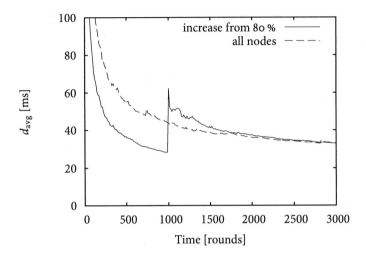

Figure 24: Effects of joining nodes. While TreeOpt works on the Meridian network, 20 % nodes join the network late in round 1000. The average
communication cost is compared to a TreeOpt run where all nodes
are present from the start of the optimization.

Drop/Join Rate	Repair Time
1 %	5 rounds
5 %	33 rounds
10 %	28 rounds
20 %	39 rounds

Table 10: Analysis of combined leaving and joining of 1 %, 5 %, 10 % or 20 % of
the nodes in the Meridian network. The time needed to repair the
tree is shown in numbers of rounds.

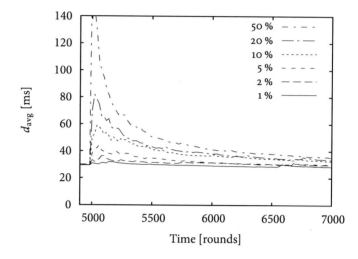

Figure 25: Analysis of combined leaving and joining. While TreeOpt works on the Meridian network, 1 %, 5 %, 10 % or 20 % of the nodes are exchanged with new nodes in round 5000. The average communication cost is shown over time while TreeOpt re-optimizes the repaired tree.

erroneous RTT measurements

urement of the RTT between any two nodes to simulate noise. While TreeOpt uses these erroneous measurements, the average communication cost was calculated using the correct network delays. The table shows the excess over the best SPT bound for the resulting trees after 10 000 rounds. For errors up to 2 % the resulting trees do not differ much from the tree TreeOpt produces without noise. The difference is within the statistical spread of the TreeOpt algorithm. An error of 5 % reduces the quality of the tree slightly. At about 5 % over the best SPT it is still quite good. An error of 10 %, which might occur in real applications, still gives acceptable solutions. An error of 20 % or higher results in a significantly worse tree with respect to the communication costs.

Figure 26 shows how TreeOpt tries to optimize these trees. For the larger noise levels, TreeOpt quickly converges to poor quality solutions. As long as the noise keeps below 5 %, the communication costs of the

Noise level	Excess
0 %	3.92 %
1 %	3.50 %
2 %	3.66 %
5 %	5.31 %
10 %	7.58 %
20 %	12.63 %
50 %	38.70 %
100 %	145.35 %

Table 11: Influence of noise on a network with 10 000 nodes. For each noise level the excess of the communication cost of the resulting tree over the best SPT after 10 000 rounds is shown.

resulting trees are indistinguishable from the expected results of Tree-Opt.

The effects of noise can also be reduced by averaging consecutive RTT measurements. This is more important for the inner edges of the tree, which are used by many connections. However, as these edges carry a larger number of messages, these messages can also be used to measure the RTT, leading to more accurate measurements.

countermeasures

3.4.5 *Discussion*

We have presented TreeOpt, a distributed algorithm for the Minimum Routing Cost Spanning Tree (MRCST). As a direct implementation of a local search for the MRCST would rely on the cooperation of a large number of peers even in smaller P2P networks, we based the Tree-Opt algorithm on an alternative problem and used degree constraint Shortest Path Trees (SPTs) to approximate the MRCST objective. We have shown that TreeOpt is effective in optimizing the MRCST objective as long as the network does not undergo heavy changes. Compared to a local search for the MRCST it can be realized with minimum co-operation of the peers in a P2P overlay. Even if the maximum child degree for each node in the tree is limited to $k = 5$, TreeOpt is able to quickly find good topologies and considerably improve the average communication cost compared to unoptimized random trees taking

alternative problem

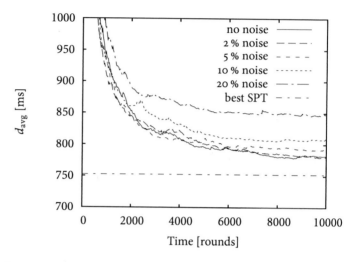

Figure 26: Influence of noise on a network with 10 000 nodes. For each noise level the communication cost of the resulting tree is shown. The best SPT gives a lower bound for TreeOpt.

only a few optimization rounds. The communication cost for these trees can be reduced by a factor of up to 8.4 for scenarios based on real world Internet measurements. We have also used randomly generated instances to demonstrate the scalability of the algorithm. TreeOpt was also shown to handle a certain amount of churn or noise.

trees vs. other topologies

Although topologies based on trees are not robust against link or node failures, TreeOpt was shown to handle leaving nodes very well, as it reconnects the tree parts within a few optimization rounds. Comparing the tree topologies against a topology based on All-Pairs-Shortest-Paths (APSPs) shows that restricting the overlay to a single tree does not exceedingly increase the average communication cost in real networks. Here, an excess of about 30 % is typical. TreeOpt and some of these results have already been published in [113].

3.5 SUMMARY

This chapter has shown global view optimization algorithms as well as a distributed algorithm for the MRCST. For the global view optimization, an evolutionary framework incorporating local search was proposed. In this Evolutionary Local Search (ELS), we used three different *global optimization* approaches for the local search. A subtree moving search was shown to be very fast but weak. An adaptation of the strong Ahuja-Murty Local Search (AMLS), which was originally designed for the Optimum Communication Spanning Tree (OCST), was shown to be very strong but slow. A cycle search was shown to be equally strong as the AMLS but much faster. We also proposed methods to increase the speed of these local searches without decreasing the solution quality too much, by introducing randomized variants of the local searches. Using any of these local searches, the ELS was shown to be more effective than a simple random restart or an Iterated Local Search (ILS).

The problem of finding an MRCST still remains a hard problem, it is $\mathcal{NP}$-hard. However, using the described algorithms, good solutions can be found very quickly, and even self-organizing networks with a minimal amount of cooperation can find approximate solutions. The *distributed algorithm* proposed distributed algorithm TreeOpt was shown to achieve good results in real world scenarios as well as random Euclidean instances. It can cope with churn and a certain degree of noise in the RTT measurements. By using degree constrained SPTs, TreeOpt was able to reduce the necessary amount of cooperation between the peers compared to a straightforward distributed version of the local searches used for global optimization.

THE SUPER-PEER SELECTION PROBLEM

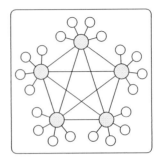

4.1 MOTIVATION

One major goal when designing Peer-to-Peer (P2P) topologies is the scalability of the P2P network. There are various approaches to increase the scalability for flat topologies by introducing regular structures, such as zones in a multi-dimensional space [147], hyper-cubes [153], prefix-routing [192, 151], or rings [167]. Another way of increasing scalability is to introduce a hierarchy within the set of peers. In these structures, some peers are appointed to be super-peers, which act as servers for the remaining peers, the so-called edge peers. Super-peers are connected among themselves, essentially forming the core of the network, while the edge peers are connected to one super-peer each [191, 98]. That way, the entire P2P network may remain connected over a considerably reduced number of links [191, 98, 80].

super-peers

Super-peer-enhanced networks do not contain a single point of failure. All P2P traffic is routed over the super-peers. This kind of P2P topology is neither fully decentralized as other P2P topologies, nor is it fully centralized. Instead, it builds a bridge between both worlds, combining the advantages of both structures. Such networks have found practical attention, e. g. by the KaZaA filesharing system [101] which builds its services on a super-peer-enhanced P2P overlay.

best of both worlds

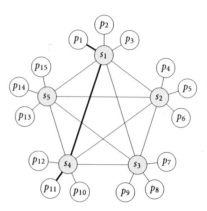

Figure 27: Example of a P2P network with selected Super-Peers. Edge peers $p_1, \ldots, p_{15}$ are connected to one super-peer each. Super-peers $s_1, \ldots, s_5$ are fully connected among themselves. The communication path between peers p_1 and p_{11} is highlighted.

example

An example of a super-peer topology with a fully meshed core is shown in Figure 27. Using a topology of this kind, any communication can be handled using only three hops. In the example, the communication between edge peers p_1 and p_{11} is routed via their respective super-peers s_1 and s_4. A broadcast in this topology can be efficiently performed by having one super-peer send the broadcast message to all other super-peers, which then forward the message to their edge peers. The super-peers can also be used to cache the status of their connected edge peers and thus further improve response times or decrease the load on the edge peers.

minimize delay

Again, we want to build topologies that minimize the average communication delay. There are two reasons for this. Firstly, we assume that peers generally wish to maintain low-delay connections to other peers. This is particularly important for applications with delay-sensitive all-to-all communication, e. g. a desktop grid. Secondly, network delay often roughly translates to network load. By using heavily congested links, an unoptimized topology would actually add to the congestion and further reduce its own usability as well as slowing down other traffic over these links. To ensure smooth operation and to ease the load on each peer, the number of super-peers as well as the number of edge peers connected to a super-peer should also be kept small.

Hence, when designing our super-peer-enhanced P2P topology, we aim for minimum communication cost as expressed by network delay in terms of Round Trip Time (RTT), and search for the optimum topology in this respect. The problem of finding the best super-peer to- *SPSP* pology, i. e. the set of super-peers and the assignment of all edge peers to these super-peers, minimizing the total communication cost for a given network is called the Super-Peer Selection Problem (SPSP).

In this chapter, we present a global view heuristic as well as two distributed algorithms for the SPSP. This chapter is structured as follows: In Section 4.2, the necessary mathematical background for the SPSP is *structure* provided. Our global view heuristic for the SPSP is presented in Section 4.3. We then present two distributed algorithms, one using a fully meshed topology for connecting the super-peers in Section 4.4, and one using a ring topology in Section 4.5. For each of the presented algorithms, we give results of experiments that were conducted on actual Internet distance measurements, demonstrating the savings in total communication cost such super-peer topologies can offer. Section 4.6 recapitulates the algorithms and findings of this chapter and presents some paths for future work.

4.2   MATHEMATICAL BACKGROUND

The SPSP is $\mathcal{NP}$-hard (see Section A.2.1 in the appendix). It may be defined as a special case of a hub location problem, first formulated by O'Kelly [131] as a Quadratic Integer Program. In this hub location *hub location* problem, a hierarchy between the nodes is established, distinguishing between so-called hubs and common nodes. This hierarchy is essentially the same as the hierarchy between super-peers and edge peers in the SPSP.

The SPSP can be formulated as an integer program. Given a network $G = (V, E)$ with $n = |V|$ nodes, p nodes are to be selected as hubs. Let x_{ik} be a binary variable denoting that node i is assigned to node k if and only if $x_{ik} = 1$. If $x_{kk} = 1$, node k is chosen as a hub. The flow volume between any two nodes $i \neq j$ is equal to one unit of flow. Since all flow is routed over the hubs, the actual weight on the inter-hub links is usually larger than one. The transportation cost of one unit of flow on the direct link between nodes i and j amounts to d_{ij}. Now, the SPSP formulated as a hub location problem is

$$\text{min: } Z = \sum_{i=1}^{n} \sum_{j=1, j \neq i}^{n} \sum_{k=1}^{n} \sum_{m=1}^{n} \left(d_{ik} + d_{km} + d_{mj} \right) \cdot x_{ik} \cdot x_{jm} \qquad (4.1)$$

quadratic
formulation

s. t.

$$x_{ij} \leq x_{jj} \qquad\qquad i, j = 1, \ldots, n \qquad (4.2)$$

$$\sum_{j=1}^{n} x_{ij} = 1 \qquad\qquad i = 1, \ldots, n \qquad (4.3)$$

$$\sum_{j=1}^{n} x_{jj} = p \qquad\qquad\qquad (4.4)$$

$$x_{ij} \in \{0, 1\} \qquad\qquad i, j = 1, \ldots, n \qquad (4.5)$$

Equation (4.1) yields the total communication cost Z. The set of constraints (4.2) ensures that nodes are assigned only to hubs, while (4.3) enforces the allocation of a node to exactly one hub. Due to constraint (4.4), exactly p hubs will be chosen in any feasible solution.

In the SPSP, we strive for $p \approx \sqrt{n}$ super-peers as a viable balance between low load and administrative efficiency. Also, the number of common peers assigned to a super-peer j is limited by

$$\frac{1}{2} \cdot p \cdot x_{jj} \leq \sum_{i=1}^{n} x_{ij} \leq 2 \cdot p \cdot x_{jj} \qquad\qquad j = 1, \ldots, n \qquad (4.6)$$

This ensures that no super-peer suffers from a high load, and that no peer is selected as super-peer without need. This constraint (4.6) is a special constraint for the SPSP, it does not occur in usual hub location problems.

A more general hub location formulation uses a demand matrix $W = (w_{ij})$. Here, w_{ij} denotes the flow from node i to j in flow units. Also, special discount factors can be applied for different edge types. Flow between hubs is subject to a discount factor $0 \leq \alpha \leq 1$, flow from a node to its hub is multiplied by a factor δ, and flow back from the hub to the node is multiplied by a factor χ. The total communication cost Z is then:

more general
problem

$$Z = \sum_{i=1}^{n} \sum_{j=1}^{n} \sum_{k=1}^{n} \sum_{m=1}^{n} \left(\chi \cdot d_{ik} + \alpha \cdot d_{km} + \delta \cdot d_{mj} \right) \cdot w_{ij} \cdot x_{ik} \cdot x_{jm} \quad (4.7)$$

In this general problem, constraint (4.6) limiting the load on a super-peer is left out.

The distinction between different types of edges is motivated by an application in the area of mail transport [51]. Here, the distribution cost differs from the collection cost. Also, the transportation cost between the hubs is assumed to be lower since more efficient means of transport can be used for the accumulated amount of flow. This extension might also be applied in the case of communication networks, especially when asymmetric links are considered. However, the most important difference from the SPSP is the introduction of demand factors w_{ij}, as will be shown in Section 4.3.3.

other applications

This general problem is known as the Uncapacitated Single Assignment p-Hub Median Problem (USAp HMP). It is uncapacitated, because it does not impose load limits on the hubs. Single assignment means that each non-hub node is assigned to one hub. The problem asks for p hubs to be selected and the average distance between all nodes is to be minimized. Other versions of hub location problems are either capacitated, use multiple assignments, or ask for the maximal distance to be minimized. Hub location problems can also be formulated allowing an arbitrary number of hubs to be created, while some additional cost for establishing a hub is introduced.

USAp HMP

Since both programs use a quadratic and non-convex objective function, (4.1) or (4.7), no efficient way to compute the minimum is known. The usual approach for solving these problems is to transform them to a Linear Programming (LP) formulation, more precisely, a linear Mixed Integer Programming (MIP) formulation. A straightforward linearization uses $\mathcal{O}(n^4)$ variables, where x_{ikjm} replaces the term $x_{ik} \cdot x_{jm}$ [20]. We will use an MIP formulation with only $\mathcal{O}(n^3)$ variables from [51, 52]:

MIP formulation

$$\text{min: } Z = \sum_{i=1}^{n} \sum_{k=1}^{n} \left(\chi \cdot O_i + \delta \cdot D_i \right) \cdot d_{ik} \cdot x_{ik} + \sum_{i=1}^{n} \sum_{k=1}^{n} \sum_{l=1}^{n} \alpha \cdot d_{kl} \cdot y_{ikl} \quad (4.8)$$

s. t. (4.2), (4.3), (4.4), (4.5),

$$\sum_{l=1}^{n} \left(y_{ikl} - y_{ilk} \right) = O_i \cdot x_{ik} - \sum_{j=1}^{n} w_{ij} \cdot x_{jk} \quad i, k = 1, \ldots, n \quad (4.9)$$

$$y_{ikl} \geq 0 \qquad\qquad\qquad i, k, l = 1, \ldots, n$$
$$(4.10)$$

flow formulation

Here, $O_i = \sum_{j=1}^{n} w_{ij}$ is the outgoing flow for node i and $D_j = \sum_{i=1}^{n} w_{ij}$ is the demand of node j. Both values can be calculated directly from the problem instance. The variables y_{ikl} denote the flow volume from hub k to hub l which has originated at peer i. Constraints (4.9) and (4.10) ensure flow conservation at each node.

An MIP formulation for the SPSP can be derived by fixing $\chi = \delta = \alpha = 1$, $w_{ij} = 1$ for $i \neq j$, $w_{ii} = 0$, and thus $O_i = D_i = n - 1$:

$$\text{min: } Z = \sum_{i=1}^{n} \sum_{k=1}^{n} 2 \cdot (n-1) \cdot d_{ik} \cdot x_{ik} + \sum_{i=1}^{n} \sum_{k=1}^{n} \sum_{l=1}^{n} d_{kl} \cdot y_{ikl} \quad (4.11)$$

s. t. (4.2), (4.3), (4.4), (4.5), (4.6), (4.10),

$$\sum_{l=1}^{n} (y_{ikl} - y_{ilk}) = (n-1) \cdot x_{ik} - \sum_{j=1, j \neq i}^{n} x_{jk} \qquad i, k = 1, \ldots, n$$

$$(4.12)$$

The factor $2 \cdot (n-1)$ for the edge-peer to super-peer links in (4.11) is the number of connections using this link. It is based on the assumption that every edge peer needs to communicate with all other $n - 1$ peers, sending and receiving messages over this super-peer link.

beware the triangle inequality violations

These formulations are equivalent to the respective quadratic programs only if the distances d_{ij} observe the triangle inequality. Otherwise, the flow model will generate solutions where messages are sent along shortest paths between two hubs instead of the intended direct link. The model can still be used for such networks. However, the resulting value can only serve as a lower bound.

lower bounds

Formulation (4.11) enables the exact solution of moderately-sized problems (up to 50 peers) in reasonable time, and additionally, the computation of lower bounds for larger networks (up to 150 peers) using its LP relaxation. For networks larger than the given threshold, we use the lower bounds described in [132]. Finally, the sum of all shortest paths' weights yields another lower bound.

4.3   GLOBAL OPTIMIZATION

In this section we present our Super-Peer Selection Heuristic, a global view heuristic for the SPSP. Our special interest lies in the construction of these P2P overlay topologies. However, as the problem of selecting the super-peers is strongly related to a hub location problem, the USA-pHMP, we also want to compare our heuristic to other heuristics for the USApHMP. The USApHMP is a well known optimization problem and has received much attention in the last two decades. With only minor adjustments, our heuristic can also be used for the USA-pHMP, which allows the comparison with other algorithms on established standard test cases.

This section is organized as follows. In Section 4.3.1, we provide an overview of related work. In Section 4.3.2, we propose our Super-Peer Selection Heuristic, and show how it can be modified for the USA-pHMP in Section 4.3.3. In Section 4.3.4, we present results from experiments on real world Internet data for the Super-Peer Selection Problem, as well as on standard test cases for the USApHMP, and compare the results with those of other established as well as recently published algorithms. Section 4.3.5 summarizes our findings.

4.3.1 Related Work

When we started our work on the SPSP, this special case of the USA-pHMP had not been studied before. However, algorithms designed for the USApHMP can also be used for the SPSP. The USApHMP is a well studied hub location problem and many heuristics as well as algorithms to calculate lower bounds have been proposed. The USApHMP was presented by Morton E. O'Kelly in [131], along with a set of test cases, called CAB. Later, O'Kelly *et al.* also presented methods of computing lower bounds for these problems [132]. *USApHMP*

Exact solutions for small USApHMP instances with up to 50 nodes have been computed by Andreas T. Ernst and Mohan Krishnamoorthy in [51]. They also present a Simulated Annealing heuristic that found good solutions for problems with up to 200 nodes. In this paper, Ernst and Krishnamoorthy also introduced a new test set, called AP. *exact solutions* *Simulated* *Annealing*

Jamie Ebery presented two more efficient Mixed Integer Programming (MIP) formulations for the special case of only 2 or 3 hubs [50], and thus solved a problem with 200 nodes (2 hubs), and a problem with 140 nodes (3 hubs). *MIP for 2 or 3 hubs*

Darko and Jadranka Skorin-Kapov presented TabuHub in [163]. This heuristic is based on Tabu Search. Two local searches were used in TabuHub: One local search tries to exchange hub nodes with non-hub nodes, the second local search tries to improve the allocation of non-hub nodes. The search is restarted multiple times from a solution where p nodes with the largest incoming and outgoing flow are selected as *TabuHub* hubs and the remaining nodes are connected to their closest hub. Results were presented only for the small CAB set, with up to 25 nodes.

Neural Network

Another type of heuristics proposed for the USApHMP are Neural Networks. In [45], Enrique Domínguez *et al.* reduced the memory consumption and the CPU time for these approaches. However, the neural network approach was again only applied to the smallest problems in the CAB set ($n \leq 15$). Unfortunately, no computation times were given, making comparisons to other heuristics difficult.

VNS-Path Relinking hybrid

The more promising heuristics for the USApHMP are based on Variable Neighbourhood Search (VNS). One example has been presented by Melquíades Pérez *et al.* in [141]. It is a hybrid heuristic combining Path Relinking [140] and VNS. This heuristic has proven to be very fast with both the CAB and AP sets, faster than any other heuristic. However, it failed to find the optimum in some of the smaller CAB instances and still left room for improvements in the larger instances of the AP set. The local search neighbourhoods used by Pérez *et al.* differ from the ones used in this thesis. Especially, the most expensive neighbourhood is missing in [141]. This explains the speed as well as the loss of quality.

GAHUB1 + GAHUB2

Two Genetic Algorithms (GAs) have been presented by Jozef Kratica *et al.* [92]. These GAs are based on different representations and genetic operations. Both feature a mutation operator that favours the assignment of peers to closer super-peers, as well as a sophisticated recombination. The results of the second GA are the best results so far, as they improved the solutions for the larger AP instances found in [141]. However, this approach did not include a local search, and can still be improved.

HGA1 + HGA2 for the SPSP

Kratica *et al.* also applied their GAs to the SPSP in [93]. In these GAs, they included the local search procedures we presented in [181]. They included caching to avoid recalculation of objective values for revisited solutions. Their GAs use a population of 150 individuals, an elitist replacement strategy, and also remove duplicated individuals from each generation and limit the number of individuals with the same objective value to ensure diversity. Using these GAs, Kratica *et al.* were able to improve the best known solutions for our SPSP instances as published in [181], while roughly triplicating the necessary CPU time.

Another recent work that needs to be included in this short survey is the VNS presented by Aleksandar Ilić *et al.* [75]. This work is accepted for publication, but the publication date is still unknown. Ilić *et al.* present two General VNS heuristics for the USApHMP and the SPSP.

They used two local searches which were developed independently but otherwise are the same as two of our local searches: *replace* and *reassign*. They also introduce a stronger third local search: *locate*. This third local search replaces a hub with a node that is not assigned to this hub. Using these local searches in their two VNSs, they improve upon the results of Kratica *et al.* for the SPSP, and also find a new best solution for the largest USApHMP instance. We will show in Section 4.3.4 that most of their results for the SPSP can still be improved.

GVNS

4.3.2   Super-Peer Selection Heuristic

The Super-Peer Selection Heuristic presented in this section is based on Evolutionary Algorithms (EAs) and local search. It operates on a global view of the network. The heuristic is quite similar to Iterated Local Search (ILS) [103], but it uses more than one offspring solution in each generation. As it combines EA techniques with local search, it can be categorized as an Evolutionary Local Search (ELS). The general work flow of the ELS is shown in Figure 28. The ELS works by starting from a valid topology and applying small changes, called moves, to the current topology. In order to escape from local optima, a mutation step is applied. A similar ELS has been applied to the Minimum Routing Cost Spanning Tree (MRCST) in Section 3.3.2.

Representation

A solution is represented by the assignment vector s. For each peer i the value $s(i)$ represents the super-peer of i. All super-peers, the set of which will be denoted by C, are assumed to be assigned to themselves, i.e. $\forall i \in C : s(i) = i$. To allow faster computation, we also store the current capacities of the super-peers, i.e. the number of peers connected to the super-peer: $|E_k| = |\{i \in V : s(i) = k\}|$. This set also includes the super-peer itself: $k \in E_k$. The sets E_k are not stored explicitly, as they are already defined by the assignment vector s.

assignment vector

Initialization

The initial solution is created by randomly selecting p peers as super-peers, and assigning all remaining peers to their closest super-peer. When handling problems with missing links, this procedure is repeated if the initial set of super-peers is not fully connected.

$s \leftarrow \text{INITIALIZATION}()$
$s \leftarrow \text{LOCALSEARCH}(s)$
$\text{mut} \leftarrow n$ *set mutation rate to problem size*
for $iter = 1 \ldots \kappa$ **do**
    $s^* \leftarrow s$
    for $i = 1 \ldots \lambda$ **do**
        if $\text{mut} = n$ **then**
            $s_{temp} \leftarrow \text{INITIALIZATION}()$
        else
            $s_{temp} \leftarrow \text{MUTATE}(s, \text{mut})$
        $s_{temp} \leftarrow \text{LOCALSEARCH}(s_{temp})$
        $s^* \leftarrow \min\{s^*, s_{temp}\}$
    if $s^* < s$ **then**
        $s \leftarrow s^*$
    else
        $\text{mut} \leftarrow \text{round}(\text{mut} \cdot \alpha)$ *decrease mutation rate*
return s

Figure 28: The Evolutionary Local Search (ELS) Framework

Local Search

After each step of the EA a local search is applied to further improve the current solution. We use three different neighbourhoods: *replacing the super-peer, swapping two peers* and *reassigning a peer to another super-peer*. If a neighbourhood does not yield an improvement, the next neighbourhood is used. It has been pointed out in [75] that the way we apply our local searches strongly resembles a General VNS with Sequential Variable Neighbourhood Descent and backward shaking [122]. The general design of the local search is shown in Figure 29.

In the first neighbourhood the local search tries to replace a super-peer by one of its children. The former child becomes the new super-peer and every other peer that was connected to the old super-peer is reconnected to the new super-peer. The gain of such a move can

replace

LocalSearch(s):
repeat
 | $s' \leftarrow s$
 | improved $\leftarrow$ false
 | **for** $i \in V$ **do**
  | **if** $s' \geq s$ **then**
   | $s' \leftarrow$ TRYREPLACESUPERPEER$(i, s(i), s)$
  | **if** $s' \geq s$ **then**
   | $s' \leftarrow$ TRYSWAP(i, V, s)
  | **if** $s' \geq s$ **then**
   | $s' \leftarrow$ TRYREASSIGN(i, C, s)
 | **if** $s' < s$ **then**
  | $s \leftarrow s'$
  | improved $\leftarrow$ true
until $\neg$ *improved* ;
return s

Figure 29: Local search used in the SPSP ELS

be computed in $\mathcal{O}(n)$ time. The following formula gives the gain for replacing super-peer k with i:

$$G_{\text{replace}}(i, k) = \sum_{j \in C} 2 \cdot |E_k| \cdot |E_j| \cdot (d_{kj} - d_{ij})$$
$$+ \sum_{j \in E_k} 2 \cdot (n - 1) \cdot (d_{kj} - d_{ij}) \tag{4.13}$$

If the gain of this move is $G_{\text{replace}}(i, k) > 0$, the move is applied.

The second neighbourhood tries to exchange the assignment of two edge peers. The gain of such a move can be computed in $\mathcal{O}(1)$ time. *swap* Since all peers connected to other super-peers are considered as the exchange partner, the total time complexity for searching this neighbourhood is $\mathcal{O}(n)$. Using the same notation as before, the following formula gives the gain for swapping the assignments of peers i and j:

$$G_{\text{swap}}(i, j) = 2 \cdot (n - 1) \cdot \left(d_{i,s(i)} - d_{i,s(j)} + d_{j,s(j)} - d_{j,s(i)} \right)$$
$$\tag{4.14}$$

The move with the highest gain is applied if its gain is $G_{\text{swap}}(i, j) > 0$.

The third neighbourhood covers the reassignment of a peer to another super-peer. Here, it is important that the capacity limits of the *reassign* involved super-peers are observed. The gain of reassigning peer i to another super-peer k can be calculated in $\mathcal{O}(p)$ time:

$$G_{\text{reassign}}(i,k) = \quad 2 \cdot (n-1) \cdot (d_{i,s(i)} - d_{i,k})$$
$$+ 2 \cdot \left(|E_k| - |E_{s(i)}| + 1 \right) \cdot d_{s(i),k}$$
$$+ 2 \cdot \sum_{j \in C \smallsetminus \{k,s(i)\}} |E_j| \cdot (d_{s(i),j} - d_{k,j}) \qquad (4.15)$$

The first part of this equation gives the gain on the link between peer i and its super-peer. The second part gives the gain on the link between the old and the new super-peer. The third part gives the gain on all remaining intra-core links. Out of all super-peers only the one with the highest gain is chosen, thus yielding a total time complexity of $\mathcal{O}(p^2)$. As SPSP strives for $p = \sqrt{n}$ super-peers, this time complexity is again equivalent to $\mathcal{O}(n)$. The move is applied only if the total gain is $G_{\text{reassign}}(i,k) > 0$.

sequence These local search steps are performed for each peer i. The local search is restarted whenever an improving step was found and applied. The local search is thus repeated until no improvement for any peer i can be found, i. e. a local optimum has been reached. Since all peers $i \in V$ are considered in these moves, the time complexity for searching the whole neighbourhood is $\mathcal{O}(n^2)$.

Mutation

Since local search alone will get stuck in local optima, we use mutation to continue the search. Mutation is done by swapping two random peers. Again, the 'gain' of such a swap can be computed by (4.14). *swapping random peers* Several mutation steps are applied in each round. The number of mutations is adapted to the success rate. The algorithm starts with $\beta = n$ mutations. If no better solution is found in one generation, the muta*mutation rate adaptation* tion rate β is reduced by 20 %. In each round at least two mutations are applied. This way, the algorithm can adapt to the best mutation rate for the individual problem and for the phase of the search. It is our experience that it is favourable to search the whole search space in the beginning, but narrow the search over time, thus gradually shifting exploration to exploitation.

Population and Selection

The heuristic uses a population of only one individual. There is no need for recombination. This is mainly motivated by the high computation cost and solution quality of the local search. Using mutation and local search, m offspring solutions are created. The best solution is used as the next generation only if it yielded an improvement. This follows a $(1 + \lambda)$ selection paradigm. If there was no improvement in the m children, the mutation rate β is reduced as described before.

only one individual

Stopping Criterion

The ELS is stopped after a fixed number of iterations κ. We also tried a stopping criterion based on the number of consecutive non-improving moves, however, we decided not to use this enhancement in the parameter studies even though it reduces the average CPU time, as it exceedingly increases the variance in the measured CPU times.

fixed number of iterations

4.3.3 Adaptation for the USApHMP

The USApHMP introduces weights w_{ij} on the connections between the nodes. While these weights have been equal for all node pairs in the Super-Peer Selection Problem, this is no longer the case in the full USApHMP. The main effect on the heuristic is that we can no longer summarize the flow on the inter-hub edges as $2 \cdot |E_a| \cdot |E_b|$. The following sum has to be used, instead: $\sum_{i \in E_a} \sum_{j \in E_b} w_{ij} + w_{ji}$. This would change the time complexity for calculating the cost of an inter-hub edge from $\mathcal{O}(1)$ to $\mathcal{O}(n^2)$. With the use of efficient data structures, however, the calculation for the cost of a move can be achieved in $\mathcal{O}(n)$ time.

Data structures

In addition to the current capacities of the super-peers we also store the weights on the inter-hub links. The value $WC(a, b) = \sum_{i \in E_a} \sum_{j \in E_b} w_{ij}$ denotes the weight on the link from super-peer a to super-peer b. In each move made by the local search or the mutation, these weights are

inter-hub link weights

changed accordingly. Only the selection of a new super-peer does not change these weights. Also, the gain calculations have to be adapted:

$$G_{\text{replace}}(i, k) = \alpha \cdot \sum_{j \in C} \left(WC(j, k) + WC(k, j) \right) \cdot \left(d_{kj} - d_{ij} \right)$$
$$+ \sum_{j \in E_k} \left(\chi \cdot O_j + \delta \cdot D_j \right) \cdot \left(d_{kj} - d_{ij} \right) \qquad (4.16)$$

$$G_{\text{swap}}(i, j) = \quad \alpha \cdot \sum_{x \in V} d_{s(j),s(x)} \cdot \left(w_{jx} - w_{ix} + w_{xj} - w_{xi} \right)$$
$$- \alpha \cdot \sum_{x \in V} d_{s(i),s(x)} \cdot \left(w_{jx} - w_{ix} + w_{xj} - w_{xi} \right)$$
$$+ \left(\chi \cdot O_j + \delta \cdot D_j \right) \cdot \left(d_{j,s(j)} - d_{j,s(i)} \right)$$
$$+ \left(\chi \cdot O_i + \delta \cdot D_i \right) \cdot \left(d_{i,s(i)} - d_{i,s(j)} \right)$$
$$+ 2 \cdot \alpha \cdot d_{s(i),s(j)} \cdot \left(w_{ii} - w_{ij} + w_{jj} - w_{ji} \right) \quad (4.17)$$

$$G_{\text{reassign}}(i, k) = \alpha \cdot \sum_{x \in V} \left(d_{s(i),s(x)} - d_{k,s(x)} \right) \cdot \left(w_{ix} + w_{xi} \right)$$
$$+ \left(\chi \cdot O_i + \delta \cdot D_i \right) \cdot \left(d_{i,s(i)} - d_{ik} \right)$$
$$+ 2 \cdot \alpha \cdot d_{k,s(i)} \cdot w_{ii} \qquad (4.18)$$

The time complexity of calculating the gains is still $\mathcal{O}(n)$ for replacing the super-peer, but has increased to $\mathcal{O}(n)$ for swapping the assignments of two peers and to $\mathcal{O}(n)$ for reassigning a peer to another super-peer. *time complexity* Using the same local search as presented for the SPSP would mean to increase the total time complexity. We therefore implemented a reduced version of the most expensive local search: the swapping of two peers. Instead of calculating the gain for swapping one peer with all other $n - p - 1$ peers, we only calculate this gain for a random sample *random sampling* of p peers, which proved to be a good compromise between computation time and solution quality.

Don't look markers

To further speed up the computation we use special Don't Look Markers (DLMs) to guide the local search. These markers are a generalization of Don't Look Bits (DLBs) that have been successfully applied for example to the Travelling Salesman Problem (TSP) [14]. In our algorithm, nodes that do not yield an improvement during one step of the local search will be marked. Nodes that are marked twice or more will not be checked in the following local search steps. However, if a node is part of a successful move all marks are removed again. This can happen if the node is the exchange partner of another node.

Using simple DLBs, i.e. disregarding all nodes with one or more marks, leads to poor results, indeed. Here, too large parts of the considered neighbourhoods will be hidden from the local search. Trying to reduce the negative impact of the DLBs immediately leads to the more general DLMs.

4.3.4   *Experiments*

We have performed several experiments to study the effectiveness of the ELS. For these experiments, we again used the real world node-to-node delay information from PlanetLab as described in Section 3.3.3. Common properties of all those networks are the frequent triangle inequality violations and missing links. Missing links are not used by the ELS when constructing or improving a super-peer topology. This means that an edge peer is not allowed to connect to a super-peer when the corresponding link is missing, and two peers cannot be allocated as super-peers at the same time when they cannot reach each other.   *PlanetLab instances*

For the USA*p*HMP we used the established CAB and AP data sets. The CAB data set from [131, 133] contains problems from the Civil Aeronautics Board. These problem instances contain up to 25 nodes, corresponding to cities in the United States. The AP data set from [51] contains data from the Australia Post. It provides problem instances with up to 200 nodes, corresponding to postal districts in the city of Melbourne. Both data sets contain full demand matrices and instructions on how to create smaller problems by either amalgamating or removing nodes. The AP data set gives two-dimensional Euclidean coordinates for the nodes, while the CAB data set gives a full symmetrical distance matrix. However, two-dimensional Euclidean coordinates for the the nodes of the CAB data set can easily be calculated from this distance matrix. In the AP set the discount factors are fixed to $\alpha = 0.75$, $\chi = 3$ and $\delta = 2$. The CAB set fixes $\chi = \delta = 1$, but provides instances for different values of α: 0.2, 0.4, 0.6, 0.8 and 1.0. When referring to the individual instances in these sets, we will use the following notations: AP.n.p for the AP instance with n nodes and p hubs, CAB.n.p.α for the CAB instance with the corresponding configuration, and shorter CAB.n.p when $\alpha = 1.0$. Both data sets have been widely accepted as   *CAB and AP data sets*

Network	Size	p	APSP	LB1	CPlex-LB
01-2005	127	12	2 447 260	2 501 413	2 632 989
02-2005	321	19	15 868 464	16 200 776	
03-2005	324	18	17 164 734	17 580 646	
04-2005	70	9	663 016	690 888	731 136
05-2005	374	20	17 324 082	17 794 876	
06-2005	365	20	18 262 984	18 735 944	
07-2005	380	20	24 867 734	25 337 991	
08-2005	402	21	27 640 136	28 151 142	
09-2005	419	21	23 195 568	23 646 759	
10-2005	414	21	28 348 840	28 905 539	
11-2005	407	21	23 694 130	24 196 510	
12-2005	414	21	20 349 436	20 885 442	

Table 12: Lower bounds for the PlanetLab instances. For each instance, the network size n, the number of super-peers p, the APSP lower bound, the LB1 lower bound, and the lower bound produced by CPlex are shown.

standard test cases, and optimal solutions for all smaller instances are known.

setup   For each problem instance the ELS was started 30 times and average values are used for the discussion. All CPU times reported in this section refer to a 2.5 GHz Xeon E5420 running Linux. The algorithms were implemented in C++.

Lower bounds

For networks too large to be handled with the models described in Section 4.2, we are interested in computing lower bounds for the SPSP. Table 12 contains the lower bounds for the networks considered here.

The All-Pairs-Shortest-Path (APSP) lower bound gives the total communication cost (i. e. the sum of the distances of all node pairs) when communication is routed over shortest paths only. Column LB1 holds a tighter lower bound as defined by O'Kelly *et al.* in [132].

CPlex bounds   The column CPlex-LB holds an improved lower bound. We let the commercial MIP solver CPlex 10.1 [76] try to solve the SPSP. CPlex essentially follows a Branch and Bound algorithm, during which both an upper and a lower bound is improved. Each feasible solution found serves as an upper bound. LP-relaxations of the current part of the search space provide lower bounds. Once the gap between upper and lower bound is reduced to 0 %, the upper bound is proven to be optimal. For the smallest network 04-2005, CPlex required over twenty-two days to arrive at a gap of 1.19 %. Due to the excessive memory usage (more than 20 GB), we stopped CPlex at that point. For network 01-2005, we stopped CPlex after thirty-two days. The value shown in the table is the lower bound at that time, but CPlex still had not found a feasible solution. For all other networks, CPlex required an exceedingly long period of time, hence was unable to provide viable results.

When compared to the lower bounds provided by CPlex, the lower bounds APSP and LB1 show their rather poor quality. This is not surprising for the APSP lower bound, as here the shortest path is used for every communication, instead of a three-hop connection in a super-peer network, where every communication is routed over the super-peers. The lower bound LB1 was designed for fast computation. It also defines an MIP formulation similar to the SPSP, but replaces the three-hop connection between edge peers with a two-hop connection. In this model, edge peers are assigned to a super-peer, but this link is not used for outgoing messages. Instead, such messages are sent directly to the super-peer of the other edge peer. Even though LB1 models more aspects of the super-peer topology than the APSP, the resulting lower bounds are still closer to the APSP lower bound than to the optimal SPSP solutions, as 01-2005 and 04-2005 show.

quality of lower bounds

Results for the Super-Peer Selection Problem

In Table 13, the best solutions ever found by our heuristic are compared with the lower bounds from the previous table. There is still a considerable gap between the LB1 lower bounds and these best known solutions, ranging from 7.1 % to 44.3 %. We believe these best known solutions to be close to the optimum, though, since the LB1 lower bound is too low in instances with many triangle inequality violations. In fact, the best known solutions for both 04-2005 and 01-2005 are still within the lower and upper bounds found by CPlex.

best known solutions

The comparison of these best known solutions with unoptimized super-peer topologies quantifies the benefit of optimization. Overlay topologies that are constructed without locality awareness can be assumed to be random. Thus, we constructed 10 000 random topologies

optimization gain

Network	Best known	Excess over LB1	Gain
01-2005	2 927 946	17.1 %	3.8
02-2005	18 584 518	14.7 %	2.8
03-2005	20 557 800	16.9 %	2.8
04-2005	739 954	7.1 %	2.6
05-2005	25 662 044	44.2 %	2.9
06-2005	22 179 526	18.4 %	3.1
07-2005	30 955 224	22.2 %	3.1
08-2005	30 866 406	9.6 %	3.2
09-2005	32 917 546	39.2 %	3.1
10-2005	32 903 884	13.8 %	3.3
11-2005	27 832 266	15.0 %	3.3
12-2005	28 428 346	36.1 %	6.9
average:		21.2 %	3.4

Table 13: Best known solutions for the SPSP. For each PlanetLab network, the cost of the best known solution, the excess over the lower bound LB1 and the gain compared to random configurations are shown.

by picking p peers to be super-peers and assigning the remaining peers to random super-peers. The column Gain in Table 13 gives the quotient of the average cost of these random configurations to the cost of the best known solution. Accordingly, the communication cost in unoptimized real world networks can be reduced by a factor of 2.6 to 3.8, and even the total communication cost for the smallest network can still be successfully reduced by a factor of 2.6 compared to unoptimized configurations. The high gain of 6.9 in network 12-2005 is a result of the large extent of triangle inequality violations in this network.

ELS results　　　The actual results of the ELS with different parameters κ and λ are shown in Tables 14 and 15. The first table shows the average excess over the best known solutions, averaged over all runs and all instances. The second table shows the corresponding average CPU times.

gaps　　　Even with the weakest parameter settings $(\kappa, \lambda) = (10, 10)$, the ELS is able to find solutions with less than 3 % higher costs than the best known solutions. Experiments with $\lambda \geq 10000$ have been stopped be-

λ	Excess over best known solution [%]						
	$\kappa = 10$	$\kappa = 20$	$\kappa = 50$	$\kappa = 100$	$\kappa = 200$	$\kappa = 500$	$\kappa = 1000$
10	2.59	1.85	1.63	1.59	1.55	1.48	1.41
20	2.08	1.43	1.20	1.18	1.16	1.11	1.06
50	1.67	1.14	0.96	0.95	0.94	0.91	0.87
100	1.40	0.95	0.77	0.76	0.74	0.72	0.69
200	1.26	0.81	0.65	0.64	0.63	0.61	0.58
500	1.03	0.63	0.50	0.49	0.48	0.46	0.44
1000	0.90	0.54	0.41	0.40	0.38	0.36	0.35
2000	0.76	0.43	0.31	0.30	0.29	0.28	0.27
5000	0.64	0.36	0.25	0.25	0.24	0.24	0.23
10000	0.53	0.29	0.20	0.19	0.18		
20000	0.44	0.24	0.16	0.16			

Table 14: Average excess over best known solutions over all PlanetLab instances for different number of offspring λ and iterations κ

λ	CPU time [s]						
	$\kappa = 10$	$\kappa = 20$	$\kappa = 50$	$\kappa = 100$	$\kappa = 200$	$\kappa = 500$	$\kappa = 1000$
10	6	11	19	29	49	110	212
20	12	22	39	60	101	223	426
50	31	56	102	154	256	563	1 077
100	63	113	210	314	520	1 141	2 179
200	126	229	427	634	1 046	2 285	4 358
500	316	586	1 095	1 624	2 677	5 845	11 132
1000	635	1 184	2 221	3 287	5 413	11 797	22 451
2000	1 273	2 383	4 462	6 597	10 855	23 659	44 999
5000	3 199	6 035	11 333	16 705	27 430	59 600	113 250
10000	6 404	12 149	22 675	33 448	54 947		
20000	12 822	24 298	45 075	66 723			

Table 15: Average CPU times over all PlanetLab instances for different number of offspring λ and iterations κ

fore $\kappa = 1000$, due to their excessive running times. For $\kappa \geq 50$ iterations the following can be observed: Increasing the number of offspring λ gives better results than increasing the number iterations κ by the same factor, using roughly the same time. For $\kappa \leq 20$ iterations and $\lambda \geq 100$ offspring, the opposite can be observed. Thus, using $\kappa = 50$ iterations leads to the best results.

CPU time
    The computation times are roughly proportional to $\kappa^{0.756} \cdot \lambda^{1.016}$. Due to the mutation rate adaptation, the ELS uses less time for each iteration κ. Since a higher number of offspring λ helps finding better solutions during the first iterations when the ELS is still using random restart, the mutation rate is reduced at a slower rate, which increases the overall runtime.

realistic settings
    For a realistic setting (i. e. runtimes well below 1000 s), the ELS still finds good solutions. Table 16 shows the results and running times for $\kappa = 50$ iterations and $\lambda = 100$ offspring. Even though the ELS only found the best solutions for the smallest instance 04-2005 using these parameters, the average gaps to the best known solutions are quite small. Only in two instances they exceed 1 %. The CPU times depend on the size of the network, ranging from less than three seconds to about five minutes for the largest instances.

best solutions
    The best solutions shown in Table 12 have been found in runs with offspring $\lambda \geq 500$. Unfortunately, neither CPlex [76], nor another heuristic, nor a good lower bound can confirm the optimality or the quality of these solutions. But, instead of attempting to prove the quality of the ELS for the SPSP, we can indeed compare it to other heuristics using the very well studied USApHMP instances.

Aleksandar Ilić
    In their yet unpublished work [75], Ilić *et al.* also experimented with these PlanetLab instances. Their VNS approach was able to find good solutions, however, in 7 of 12 instances, we found better solutions. A comparison between our and their best solutions is shown in Table 17.

Results for the USApHMP

Since the full USApHMP is more complex than the SPSP, we used the adapted ELS as described in Section 4.3.3 for these experiments. The AP set consists of 20 smaller ($n \leq 50$) and 8 larger instances ($n \geq 100$). Optimal solutions are known only for the smaller instances. The CAB set consists of 60 instances (four different sizes up to $n = 25$ nodes, three different numbers of hubs, five different discount factors α). For

Network	Average cost	Excess over best known	CPU time per run	# best found
01-2005	2 937 991	0.34 %	16.2 s	0/30
02-2005	18 722 844	0.74 %	155.1 s	0/30
03-2005	20 765 516	1.01 %	156.1 s	0/30
04-2005	740 927	0.13 %	2.9 s	5/30
05-2005	25 956 634	1.15 %	241.0 s	0/30
06-2005	22 388 734	0.94 %	232.3 s	0/30
07-2005	31 149 618	0.63 %	240.6 s	0/30
08-2005	31 157 564	0.94 %	274.8 s	0/30
09-2005	33 212 408	0.90 %	308.5 s	0/30
10-2005	33 198 120	0.89 %	295.5 s	0/30
11-2005	28 047 932	0.77 %	293.1 s	0/30
12-2005	28 650 124	0.78 %	304.4 s	0/30

Table 16: Detailed results for the PlanetLab instances using κ = 50 and λ = 100. For each PlanetLab instance, the average cost, the gap to the best known solution, the CPU time, and the number of runs that found the best known solution are shown.

all these instances the optimum is known. Again, each experiment was repeated 30 times.

Using only a small number of offspring λ = 50, the ELS already found the optimal solutions for the smaller instances of the AP set in every run after at most κ = 10 iterations, taking 0.12 s (or after κ = 5 iterations, taking 0.10 s without DLMs). A similar result can be observed for the CAB set. Here, all runs with λ = 100 reached the optimal solutions after κ = 10 iterations, taking 0.03 s with (or 0.05 s without DLMs). *smaller instances*

This excellent solution quality can not be kept up for the larger instances of the AP set. Using realistic parameters (κ = 20, λ = 100), the ELS fails to find the best known solutions in some runs. Table 18 shows the results. Only in 63 out of the 240 runs on these instances the best known solution was reached. However, the average excess above those best known solutions is still considerably good, ranging from 0.00 % for AP.100.5 to 1.32 % for AP.200.20. *larger instances*

Network	Wolf	cmp	Ilić	gap
01-2005	2 927 946	=	2 927 946	0 %
02-2005	18 584 518	>	18 579 238	0.028 %
03-2005	20 557 800	<	20 569 390	−0.056 %
04-2005	739 954	=	739 954	0 %
05-2005	25 662 044	<	25 696 352	−0.134 %
06-2005	22 179 526	<	22 214 156	−0.156 %
07-2005	30 955 224	<	30 984 986	−0.096 %
08-2005	30 866 406	<	30 878 576	−0.039 %
09-2005	32 917 546	<	32 959 078	−0.126 %
10-2005	32 903 884	>	32 836 162	0.206 %
11-2005	27 832 266	>	27 787 880	0.160 %
12-2005	28 428 346	<	28 462 348	−0.119 %

Table 17: Comparison of best known results. The best results found by the ELS are compared to the best results found by the VNS from Ilić *et al.* in [75]. The column 'cmp' shows which results produced less costs, and the column 'gap' gives the percentage excess of the ELS solutions over the VNS solutions.

Instance n p	CPU time	Excess over best known solution	Best known cost	# best found
100 5	6.8 s	0.00 %	136 929. 444	30/30
100 10	14.6 s	0.21 %	106 469. 566	17/30
100 15	23.3 s	0.61 %	90 533. 523	2/30
100 20	29.6 s	1.25 %	80 270. 962	2/30
200 5	29.2 s	0.17 %	140 062. 647	11/30
200 10	70.2 s	0.16 %	110 147. 657	0/30
200 15	104.6 s	0.62 %	94 459. 201	1/30
200 20	145.6 s	1.32 %	84 955. 367	0/30

Table 18: Results for the larger AP instances ($n \geq 100$) using $\kappa = 20$ and $\lambda = 100$. For each instance, the average CPU time, the excess over the best known solution, the cost of the best known solution, and the number of runs that found the best known solution are shown.

Instance		CPU	Excess over best	Best known	# best
n	p	time	known solution	cost	found
100	5	6.8 s	0.00 %	136 929. 444	30/30
100	10	16.3 s	0.15 %	106 469. 566	21/30
100	15	25.7 s	0.44 %	90 533. 523	6/30
100	20	32.0 s	1.20 %	80 270. 962	3/30
200	5	37.1 s	0.02 %	140 062. 647	27/30
200	10	84.9 s	0.17 %	110 147. 657	2/30
200	15	132.8 s	0.51 %	94 459. 201	3/30
200	20	177.2 s	0.98 %	84 955. 367	0/30

Table 19: Results for the larger AP instances ($n \geq 100$) using $\kappa = 20$ and $\lambda = 50$, without Don't Look Markers. For each instance, the average CPU time, the excess over the best known solution, the cost of the best known solution, and the number of runs that found the best known solution are shown.

The DLMs described in Section 4.3.3 help to speed up the ELS by a factor of about 2. However, they also hide some parts of the search space from the local search. A comparison between the ELS with and without DLMs in Table 19 shows that an ELS without DLMs using only half the number of offspring λ obtains slightly better results in roughly the same amount of time.

don't look markers

The best known solution for AP.200.20 has been found in five out of all 210 runs of the ELS without DLMs. In one run using $\lambda = 20$, the ELS found this solution in the tenth iteration after only 36 s. Other runs that found this solution used $\lambda = 100$, $\lambda = 500$, and two runs with $\lambda = 1000$, taking between 150 s and 30 minutes and between six and nine iterations. This solution improves previously published results by 0.20 %. All other best known solutions for the larger instances of the AP set have been listed in [92], with the exception of AP.200.5 and AP.200.15 which have already been improved by our ELS in [181]. The ELS is able to reach these best known solutions while taking roughly the same CPU time as the GAs presented in [92]. In their yet unpublished work, Ilić *et al.* further improve our solution for AP.200.20 by 0.000 046 % [75].

best known solutions

number of hubs

The ELS seems to be more effective in finding the best solutions for problems with less hubs. For example it never failed to find the best solution in AP.100.5 and also finds the best solution for AP.200.5 in nine out of ten runs. For problems with more hubs the success rate decreases and the average excess over the best known solution increases. Also, since the search is more complex for problems with more hubs, the

CPU time

average CPU time increases with the number of hubs. The computation times range from well below a second for all smaller instances to about three minutes for the largest instance in the AP set. The computation times for the CAB set seem independent on the intra-core discount factor α. The larger the problem instance and the more hubs are to be located the more time the local searches use. This behaviour can also be observed for other heuristics and is therefore not surprising.

For all runs, we observe that the ELS seems to converge around iter-

best parameters

ation $\kappa = 20$, independent of the use of DLMs. Although it still finds some improvements till $\kappa = 1000$, the additional CPU time is not well spent. Instead of increasing the number of iterations beyond $\kappa = 20$, it is better to increase the number of offspring λ. Reducing the number of iterations below $\kappa = 20$ also produces worse results. In such settings, the ELS works too much like random restart and fails to exploit the structure of the search space.

4.3.5   Discussion

USApHMP

In this section, we have presented a hybrid heuristic combining evolutionary algorithms and local search for the Super-Peer Selection Problem (SPSP) and the related Uncapacitated Single Assignment p-Hub Median Problem (USApHMP). This Evolutionary Local Search (ELS) has proven to find optimal solutions in all USApHMP instances where an optimum is known, and to find or even improve the best known solutions for the larger USApHMP instances.

SPSP

The time complexity of the local searches improves for the SPSP. Here, the ELS is able to handle problems with $n = 400$ and more nodes within reasonable time. Since we want to find the best topologies to be used as a comparison for the distributed algorithms presented in the next section, we also let the ELS search solutions using larger than reasonable parameters (up to $\lambda = 10000$ offspring). Based on the excellent results for the USApHMP, we can conclude that the best known solutions found during these excessive runs are very close to the still unknown optimal solutions.

Even with more relaxed parameters, the ELS has been shown to find solutions for the SPSP with a very small gap the the best known solutions. In comparison to unoptimized topologies, the total communication costs in real world super-peer networks can be reduced by a factor of about 3.

gain

4.4 DISTRIBUTED ALGORITHM: SPSA

The Super-Peer Selection Algorithm (SPSA) presented in this section creates a P2P topology in which some peers will act as super-peers while the remaining peers, called edge peers, are assigned to exactly one of the super-peers each. In this topology, the super-peers are expected to fulfil certain management tasks, mainly forwarding messages. They are fully connected among themselves, so they can interact directly with their edge peers and with their fellow super-peers. Edge peers route all communication via their respective super-peer.

super-peer-enhanced overlay

The SPSA aims at creating such a topology with minimum communication cost. In a P2P setting, this cost can be thought of as the average end-to-end network delay between any two hosts. This helps reducing response times for queries in a file-sharing application, but also is of importance in desktop grids, multimedia applications, real-time online games, and many other P2P systems. As a side effect, the reduced communication cost also translates to a reduced load in the underlying network.

minimize communication delay

In order to solve the problem of selecting the best set of super-peers in a distributed, scalable and self-organized way, SPSA not only creates a super-peer-enhanced P2P topology, but also enables the peers to apply local moves to improve this topology. These local moves are inspired by the global optimization presented in the previous section, but they need to take into account that peers do not have full knowledge about the network. Firstly, a P2P network can change over time due to leaving and joining nodes, such that even the number of participating peers is not known to each peer. Secondly, the RTT on each link can change due to routing decisions, congestions, or simply increased load. Thirdly, measuring all RTTs in the P2P network also puts a high load on the network and has to be avoided in a self-organizing system that aims at reducing the total communication cost.

local moves

This section is organized as follows. In the remainder of this section, the necessary additional mathematical background is introduced. In Section 4.4.1, we provide an overview of related work. In Section 4.4.2, we present the Super-Peer Selection Algorithm (SPSA). As SPSA heavily depends on RTT values, we describe several methods for generating RTT estimations based on network coordinates in Section 4.4.3. In Section 4.4.4, the simulation framework used to conduct experiments on real world Internet networks is presented. The results of these experiments are shown in Section 4.4.5, demonstrating the savings in communication cost SPSA can achieve. The section is concluded and paths for future work are laid in Section 4.4.6.

Additional Mathematical Background

The problem SPSA tries to solve is the SPSP from Section 4.2. However, when considering P2P networks, the total communication cost Z from (4.1) is often meaningless, as it depends on the size n of the network. Instead, the average communication cost d_{avg} can be computed as $d_{avg} = Z \cdot \frac{1}{n \cdot (n-1)}$. This average cost gives the expected RTT for any communication in the topology. It can be compared directly to usual RTTs in other networks.

average RTT

In symmetric networks, the cost function (4.1) can be rewritten. The set of super-peers C makes up the *core* of the topology. Links from edge peers to their super-peers are referred to as *spokes*. Denoting the set of peers assigned to super-peer k including k itself as E_k, the total communication cost Z can be written as:

$$Z = 2 \cdot (n-1) \cdot \underbrace{\sum_{i,k} d_{ik} \cdot x_{ik}}_{\text{spoke connections}} + \underbrace{\sum_{k,m} d_{km} \cdot x_{kk} \cdot x_{mm} \cdot |E_k| \cdot |E_m|}_{\text{intracore connections}}$$

(4.19)

This formulation clearly shows the weights put on each link. Spokes are weighted by a factor of $2 \cdot (n-1)$, while intracore links are weighted by a factor of $2 \cdot |E_k| \cdot |E_m|$. Each link is used in both directions, accounting for the factor 2. For the intracore connections, this factor appears implicitly in the double sum.

edge weights

These weights become roughly equal if the number of edge peers each super-peer serves is $|E_k| \approx \sqrt{n}$. In such configurations, both types of edges are weighted by a factor of about $2 \cdot n$. In the SPSA, both types of edges will be considered equally weighted, even if the edge peer sets E_k are slightly larger or smaller than the desired value. This allows SPSA to perform calculations without the knowledge of the network size n or the number of peers $|E_k|$ another super-peer k is serving.

approximately equal

4.4.1   Related Work

Previous proposals for super-peer topology construction mechanisms include *SG-1* presented by Alberto Montresor in [125]. SG-1 builds a super-peer topology on top of an existing overlay. The super-peers are connected using a random structure. SG-1 aims for a minimum number of super-peers according to given capacity constraints. The underlying topology is used to ensure connectivity and provide means to select a random peer. In *SG-1*, the gossip protocol *Newscast* [79] is used in this underlying topology.   *SG-1*

The successor *SG-2* presented by Gian Paolo Jesi, Alberto Montresor and Özalp Babaoglu in [80] also operates on the basis of a gossip mechanism, but tries to minimize the communication cost in terms of latency. In SG-2, a maximum latency is defined. Clients are assigned to super-peers only if their latency distance is smaller than this threshold. A similar threshold is used for the connection between super-peers, and the capacity limits for the super-peers from SG-1 are used again. The authors propose the construction of this super-peer overlay topology with the aid of network coordinates. The suggested biology-inspired mechanism mimics the social behaviour of insects to promote particular nodes to super-peer level. Although SG-2 and SG-1 try to minimize the number of super-peers, the resulting topologies use $\mathcal{O}(n)$ super-peers, while SPSA creates topologies with only $\mathcal{O}(\sqrt{n})$ super-peers. Also, the connections between the super-peers are not part of the optimization. Moreover, because of the explicit latency thresholds, SG-2 does not ensure the connectivity of the overlay and needs to rely on an underlying topology to keep the network connected.   *SG-2*   *comparison to SPSA*

An approach to create structured super-peer connections is given by Michael Kleis *et al.* in [89]. The proposed lightweight super-peer topologies try to approximate a Minimum Spanning Tree (MST) by using edges from a two-dimensional Yao graph. The two-dimensional coordinates for the super-peers are created using a landmark-based scheme called Highways. In this topology, only super-peers join the Yao graph, edge peers are connected to the closest super-peer that is capable of accepting a new peer. The decision to become a super-peer is made only once when entering the overlay and no downgrade strategy is provided. As the Yao graph gives a bound on the node degree, the number of connections a super-peer has to maintain is limited. However, the network size and the number of super-peers are unrelated, and   *Yao graphs*

message routing load is unevenly distributed among the super-peers, it is higher in the centre and lower towards the edge of the Yao graph.

super-peers in use

An overview of P2P overlay schemes by Eng Keong Lua *et al.* in [104] shows that the concept of super-peers was already used in early P2P file sharing applications such as FastTrack/KaZaA or Gnutella, where they allowed the networks to scale better.

landmark based approaches

There are also some efforts related to topology-aware P2P topology construction without the distinction between edge peers and super-peers. Jawwad Shamsi *et al.* present TACON in [154], a landmark based topology construction scheme that tries to match the structure of the underlying network by minimizing the latency on the used edges. In this scheme, nodes are places in zones by measuring the distances to all landmarks and picking the closest landmark. These landmarks keep a list of all nodes in their zone and provide them with a list of close neighbours, using the measured latencies as a coordinate vector. The actual topology is constructed by using links to short jump neighbours, i.e. nodes in the same zone, and links to long jump neighbours. This approach resembles the binning method introduced by Sylvia Ratnasamy *et al.* in [148]. In this binning scheme, nodes first measure their network latency to a fixed set of landmarks and then assign themselves to virtual bins which each represent one particular ordering of the landmark latency results. Ratnasamy *et al.* also show how their scheme can be used for structured as well as unstructured networks, by providing topological hints.

4.4.2   Super-Peer Selection Algorithm

In this section, the Super-Peer Selection Algorithm (SPSA) is presented, a fully distributed algorithm to solve the SPSP. Each peer in the network runs the same algorithm and tries to improve the topology. SPSA operates in a self-organized way, enabling the topology to automatic-

failure resilience

ally recover from node failures. When a super-peer crashes or leaves the network, its edge peers will switch to another super-peer. When an edge peer crashes, its super-peer simply removes the unnecessary connection.

the super-peers' load

In order to minimize the number of connections the super-peers has to maintain, SPSA aims for $|C| = \sqrt{n}$, where n is the number of peers. Once the edge peers are equally distributed among the super-peers, every super-peer serves $|E_k| = \sqrt{n}$ peers, including itself, and maintains links to the remaining super-peers. This value minimizes

every γ time units do:
$i \leftarrow \text{NEARESTSUPERPEER}()$
if $s(i) \neq i$ **then**
    $\text{CONNECTTO}(i)$
    $s(i) \leftarrow i$

Figure 30: SPSA, edge peer part. In each round, the edge peer tries to connect to a closer super-peer.

the number of links for each super-peer. Every other value for $|C|$ will increase the expected number of links for the super-peer: For smaller values the number of spokes increases, and for larger values the number of intracore connections increases, in both cases outweighing the respective savings.

Edge peers maintain only one link: the link to their super-peer. In SPSA, each edge peer tries to establish a link to the closest super-peer. Thus, it minimizes the network delay for this link, and since every com- *minimal spokes* munication is routed over the super-peer, it also reduces its distance to all other peers in the network.

Due to this greedy behaviour of the edge peers, a super-peer might find itself serving a large number of edge peers. In SPSA this number is restricted to a certain degree to limit the load on the super-peers. The super-peer can choose to reject new edge-peers, or appoint a new super-peer. With this restriction, SPSA achieves a reasonably even dis- *number of* tribution of the edge peers as well as an adaptation of the number of *super-peers* super-peers to the size of the network.

Rules

SPSA operates on a set of rules to improve routing efficiency. Every γ time units, each peer performs a reorganization cycle. Edge peers try to improve their spoke connection and connect to their closest super-peer *edge peer* unless they are already connected to it. They receive a list of super-peers from their current super-peer. The edge peer part of SPSA is shown in Figure 30.

If the connection to its super-peer is lost, the edge peer restores its link to the overlay core by contacting the closest of the remaining super-peers from its list.

super-peer

The super-peer reorganization step depends on the current length of the super-peer list $|C|$. One aim of SPSA is to achieve a core size of $|C| \approx \sqrt{n}$. Even though a super-peer k does not necessarily know the network size n – it only knows its edge peers E_k and the core C – this aim can be achieved quite well by comparing the number of edge peers $|E_k|$ to the number of super-peers $|C|$. In the desired setting, the number of super-peers $|C|$ in the network and the number of edge peers each super-peer serves are equal: $|C| = |E_k| = \sqrt{n}$. However, this can only be achieved under unrealistic assumptions: n must be square, and edge peers must find the right super-peer. So, SPSA uses relaxed limits.

*up- and
downgrade*

If $|E_k| > 2 \cdot |C|$, the super-peer considers its current load to be too high. To reduce the load it promotes one of its edge peers to super-peer level. If too many super-peers have been created or if the network shrinks, some super-peers may be underloaded. If $|E_k| < 1/2 \cdot |C|$, the super-peer considers its load too low, and downgrades to edge peer level.

replacement

Even if the number of edge peers remains within these bounds, the super-peer tries to find a replacement super-peer. It searches for a better super-peer $j \in E_k$, which would serve as a super-peer for all other edge peers of k and k itself. The following formula is used to find the best replacement j^*:

$$j^* = \arg\max_{j \in E_k} G(E_k, C, j, k) \qquad (4.20)$$

$$G(E_k, C, j, k) = \sum_{i \in C} d(i, k) - \sum_{i \in C} d(i, j)$$
$$+ \sum_{i \in E_k} d(i, k) - \sum_{i \in E_k} d(i, j) \qquad (4.21)$$

$G(E_k, C, j, k)$ gives the expected gain of replacing super-peer k with j, replacing the links to the core C and all edge peers E_k. This equation is an approximation of the real gain (4.13), assuming that the weights on intracore links and spokes are similar. If a super-peer candidate $j^* \neq k$ is found, it becomes a new super-peer, and all remaining edge peers from E_k including k itself switch to it. The super-peer part of SPSA is shown in Figure 31.

*number of
measurements*

Evaluating $G(E_k, C, j, k)$ means measuring the communication cost of all links from k and j to all super-peers C and all edge peers E_k. This measurement may be achieved by sending a delay measurement message across the link. As there are approximately $\sqrt{n}$ super-peers in the network running this algorithm, each super-peer needs to check $\sqrt{n}$

every γ time units do:

if $|E_k| < \frac{1}{2} \cdot |C|$ **then**
  ⌊ $\mathrm{role}_k \leftarrow$ edge-peer   *downgrade*

if $|E_k| > 2 \cdot |C|$ **then**
    $j^* \leftarrow \arg\max\limits_{j \in E_k \setminus \{k\}} G(E_k, C, j, k)$
  ⌊ $\mathrm{role}_{j^*} \leftarrow$ super-peer   *promotion*

if *else* **then**
    $j^* \leftarrow \arg\max\limits_{j \in E_k} G(E_k, C, j, k)$
    if $j^* \neq k$ **then**
        ⌈ $\mathrm{role}_{j^*} \leftarrow$ super-peer   *replacement*
        ⌊ $\mathrm{role}_k \leftarrow$ edge-peer

Figure 31: SPSA, super-peer part. In each round, the super-peer k downgrades if its load is too low, promotes an edge peer if its load is too high, or searches for a replacement super-peer if its load is neither too low nor too high.

edge peers, and $2 \cdot \sqrt{n} - 2$ links need to be measured for each candidate peer, the total number of measurements amounts to $\mathcal{O}(\sqrt{n^3})$ in every round.

Due to this large number of measurements, SPSA would be unable to scale when each measurement requires to send actual delay measurement messages across a link. However, the use of network coordinates, as described in Section 4.4.3, considerably reduces the amount of messages, allowing the algorithm to scale well.

Messages

SPSA uses a set of messages for its operation. In the bootstrapping phase, a peer i that wants to join the overlay first contacts an arbitrary *joining* node j known to belong to the overlay, and sends a QUERYSUPERPEER message to j. Peer j responds with a SUPERPEERLIST message, containing a list of all known super-peers. Peer i then picks the closest super-peer k from the list and tries to connect to k using a CONNECT message. Depending on its number of edge peers, k replies either with an AC-CEPT or a REJECT message. In the case of ACCEPT, the link between i and k is established and i is added to the edge peer set E_k. In the case

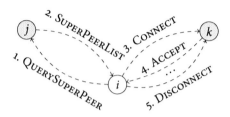

Figure 32: Edge peer life cycle. The typical message flow of an edge peer is shown from joining the overlay until leaving it.

of REJECT, *i* will pick another super-peer from the list and try again. When *i* wants to leave the network or connect to another super-peer, it sends a DISCONNECT message to *k*. The life cycle of a simple edge peer is shown in Figure 32.

leaving

A peer changes its super-peer by sending a CONNECT message to the new super-peer. The new super-peer responds with an ACCEPT or REJECT message. The edge peer still has to disconnect from the old super-peer by sending a DISCONNECT message. Super-peers keep track of their edge peers and provide them with a current list of super-peers, particularly when they issue a REJECT message to a prospective edge peer.

changing the super-peer

When a super-peer *k* determines it is supporting too many edge peers, it picks one of its edge peers *j** according to (4.20), and appoints it super-peer by sending a NEWSUPERPEER message to *j**. Peer *j** may choose to refuse this appointment, e. g. if it is unable to fulfil the requirements of a super-peer. If peer *j** accepts the appointment, it replies with an ACCEPTSUPERPEER message and broadcasts its new status by sending INSERTSUPERPEER messages to all super-peers. This process is shown in Figure 33.

creating a new super-peer

The replacement of a super-peer should be handled carefully. If the super-peer simply downgrades, its edge peers would eventually connect to another super-peer, but in the meantime the network would be split. To avoid this, the super-peer wishing to promote a replacement peer first sends a RECONNECT message to its edge peers, asking them to switch to the new super-peer voluntarily. It then waits a pre-defined time span, called the Pre-Reject Phase, and sends a REJECT message to each of the remaining edge peers, disconnecting them. It also sends a REMOVESUPERPEER message to all super-peers which in turn may for-

super-peer replacement

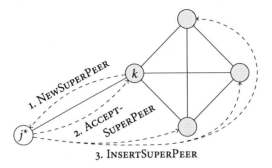

Figure 33: Appointing a new super-peer. An overloaded super-peer selects one of its edge peers to become a new super-peer. If this peer accepts, it is inserted in the core by establishing connections to every other super-peer.

ward this information to their respective edge peers. The message flow for this scenario is shown in Figure 34.

The downgrade of a super-peer is handled in the same way, but in this case the RECONNECT message does not contain a redirection to a specific super-peer. Instead a current super-peer list is provided and edge peers pick the closest super-peer from this list.   *downgrade*

4.4.3   *Network Coordinates*

The pure SPSA as presented in Section 4.4.2 would use a lot of messages to measure the RTT of the network links in order to calculate the gain of each local move in (4.21). As pointed out, this is counterproductive and needs to be avoided. Network coordinates provide the means to estimate RTTs without measuring all links in a network. Generally, network coordinates cause only a fraction of the network load of standard delay measurement methods, even when considering the additional administrative overhead. With the help of network coordinates, SPSA becomes a scalable distributed algorithm. This section describes network coordinates in general and concentrates on three different methods of generating them. A recent survey covering many more aspects of network coordinates can be found in [46]. A comparison of network coordinates to other network-aware overlay construction support methods is given in [143].

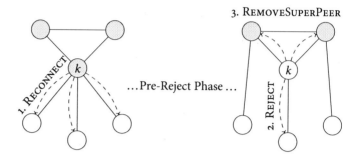

Figure 34: Super-peer replacement. Before downgrading to edge peer level, the old super-peer informs its edge peers about its replacement. All edge peers that did not switch to the new super-peer after a certain time span will be rejected when the super-peer finally downgrades.

embedding space

Network coordinates are space coordinates assigned to each node in a given network individually. The space used for network coordinates is usually a Euclidean space; sometimes a spherical space is used. The individual coordinates are carefully chosen, such that the distance between any two nodes in that space approximates the measurable delay between those nodes in the network. Thus, a node is able to estimate the delay to a remote node without sending delay measurement messages, once it has the knowledge of the other node's coordinates.

embedding error

The network coordinates assigned to each node are generated by embedding the network into a target space. As the target space is almost always a metric space, it cannot represent triangle inequality violations which occur frequently in real world networks. Also the number of dimensions of the target space should be kept low to allow coordinates to be transferred using short messages. However, the lower the dimension, the less information about the original network can be saved. For these reasons, error-free embeddings are practically unavailable [193]. Furthermore, the generation of optimal network coordinates is a continuous non-linear optimization problem, and algorithms used to embed the network may introduce an additional embedding error if they fail to locate an optimum solution.

dissemination of coordinates

When the network coordinates have been calculated and assigned to each node, the nodes still need to learn about other nodes' coordinates before estimating their distance to these nodes. The usual method to propagate the coordinates is to transmit them via background gossip or piggybacked on existing messages, but in the context of SPSA, super-peers can store the coordinates of other super-peers in the super-peer list and provide these coordinates to their edge peers. The coordinates of an edge peer are only used by its super-peer when it searches for a replacement peer. They do not need to be transferred to any other peer.

Network coordinates mechanisms can be divided in two categories: Static approaches compute coordinates for every node only once, as there is no explicit need to recompute these coordinates. In contrast, dynamic methods are based on the recomputation of the coordinates over time. Here, the initial configuration is often random, but the coordinates eventually approximate a good embedding. This difference also has influence on the message complexity – dynamic methods require an ongoing message exchange to update and to propagate the coordinates.

static vs. dynamic

Vivaldi

A successful example for the dynamic approach is Vivaldi [32, 33]. It is inspired by the effect of spring deformation. In the Vivaldi approach, when two nodes communicate, they measure both their link's delay and their spatial distance, and set up a virtual spring. Should the measured delay and spatial distance be identical, the spring rests in a zero-energy state that represents the optimum. Otherwise, the spring is expanded or compressed, and the nodes are either pulled closer towards each other or pushed apart in the embedding space. The mismatch between measured delay and spatial distance determines how far the coordinates are moved.

spring network

In a real spring network – disregarding the fact that it could easily contain too many springs to be built – the degree of expansion or compression defines the force that moves the nodes, i. e. acceleration. However, this would lead to an eternal oscillation of the spring, even in a network with only two nodes. Vivaldi does not simulate the spring network, it rather provides a distributed algorithm to minimize the degree of expansion and compression, finding an equilibrium of all forces. The degree of expansion or compression is therefore not used as a value for acceleration, instead it is used directly as the length of the movement.

the physical world

The Vivaldi coordinates can be initialized using random coordinates, or even by placing all nodes at the same point. In the latter case, the

initialization

direction of the spring is undefined, so if the spring needs to expand and push both nodes apart, the direction is chosen at random.

oscillation

Just like a spring network, Vivaldi tends to oscillate. To dampen these oscillations, a factor δ is introduced to reduce the length of movements in later iterations. According to [33], δ should be initialized to 1.0 to quickly achieve good results. The value should be reduced by 0.025 in each iteration as the quality of the network coordinates increases with each round. It should also not be reduced below 0.05, in order to allow the system to adapt to changes of the underlying network.

extensions

There are various extensions to the initial presentation of Vivaldi in [32]. In order to better represent peers that are connected to the network via a low-bandwidth connection, such as dial-up Internet connections, the concept of heights is introduced in [35]. The distance between nodes is then calculated by adding these heights to the normal spatial distance. Another possible extension is to use negative heights. This allows to introduce triangle inequality violations in the embedding space. However, the rules for adapting this negative height would have to be formulated such that they do not conflict with the rules for positive heights. As far as we know, no functional algorithm using negative heights has yet been published.

GNP

A widespread static method is Global Network Positioning (GNP) [130].

landmarks

In GNP, a set of nodes is selected to act as landmarks, serving as points of reference for other nodes that wish to embed themselves into the target space. This approach is implemented using an optimization-driven scheme. In a first step, the set of at least $d + 1$ landmark nodes is embedded in the d-dimensional target space. The second step is taken by every other node individually, as it solves an optimization problem to find its coordinates based on its distance to these landmarks. The original version of GNP uses Simplex Downhill [78, 129] to solve these optimization problems.

random landmarks

Several properties of GNP have been investigated and improved by subsequent work. GNP's original requirement of fully fixed landmarks has been relaxed in [30] by using random nodes as landmarks. This allows to distribute the load of the landmarks to a larger set of nodes.

An advanced method to solve the optimization problems of embedding the individual node has been introduced in [111]. It is based on

d-dimensional hypercubes. In each iteration, out of the 2^d subcubes *hypercubes*
the one that promises the best coordinates is chosen. The hypercube
method can also backtrack to a certain degree, as the edge length of
the subcubes is not necessarily reduced to half the size of the original
hypercube. Thus, it can also reach regions of the search space outside
of the initial hypercube. Once the edge length falls below a pre-defined
accuracy ε, the median of the final hypercube is used as the coordinate
of the node.

The embedding quality of GNP benefits immensely from both improvements. In SPSA, we will not use the original version of GNP,
but include these improvements and refer to the improved version as
GNP*. *GNP**

Geographical coordinates

An approach to produce static coordinates without a single measurement is the use of geographical coordinates. While Vivaldi or GNP base
their coordinates on actual delay measurements, implicitly incorporating link capacity and utilization, geographical approaches assume that
the physical distance strongly correlates with the RTT. The coordinates assigned to each node represent their geographical location on the *Consider a*
Earth's surface, encoded as latitude and longitude. The spatial distance *spherical Earth*
is then measured using a great circle's section running through both *...*
points.

The authors of GNP have examined the aspect of geographical coordinates in their work [130] and found GNP's distance measurement *GNP vs.*
to go beyond simple geographical relationships. RTTs in the Internet *geographical*
also depend e.g. on the availability of high bandwidth connections
or on routing policies. However, geographical coordinates can be obtained easily, even without the use of a Global Positioning System.

There are various projects and commercial services providing geographical information based on IP addresses. The authors of GNP used
data from the NetGeo project, which has stopped active maintenance
of its data before 2003. NetGeo's technology has been licenced to Ixia, *dead or*
who provided a commercial service called IxMapper, which again is *commercial*
discontinued since 2006. Another commercial alternative is Akamai's *sources*
EdgeScape. Akamai provides servers around the world and uses this
technique in combination with a network load detection system to select the best possible server for each user, thus improving response
times and avoiding congestions. In [135], three methods for mapping IP

addresses to geographical coordinates, GeoTrack, GeoPing, and Geo-Cluster, were introduced, but neither has been publicly deployed.

A public source for geographical coordinates that has no query restrictions, covers the entire IP address space, and is still maintained is GeoLite City from MaxMind, Inc. [106]. GeoLite City is a free version of the commercial GeoIP City database, which is also provided by MaxMind. As GeoLite City is free of charge, and it is advertised to be as accurate as many other commercial geolocation solutions, we used it to assign latitude and longitude information to known IP addresses in the PlanetLab instances.

public source

4.4.4   *Simulation Framework*

For the development and evaluation of SPSA, a lightweight simulation framework, called NetSim, has been developed. One design goal of NetSim was to allow the simulated protocol to be converted into a real distributed application with comparatively little effort, allowing simulation results to be verified in a real setting. The benefit of re-using code in this way has been pointed out in [128].

NetSim

NetSim is an event-based simulation framework. Its main component is a global message queue which distributes messages to individual nodes. In this queue, messages are stored according to their arrival time, which is based on the communication delay for the respective link. NetSim abstracts from lower layers and only uses the message delay for each link. The event-based simulation removes the first message from the queue and hands it to the receiving node, which can then react on the message, change its internal state or send other messages. All communication in NetSim is asynchronous. Since the communication delay is strictly observed, multiple messages for the same link can be present in the queue. The simulation is terminated when the queue is empty, after a pre-defined time span, or when a desired event has been reached.

message queue

The queue is also used to start repetitive tasks on a node or emulate a local timeout. To this end, a node generates a special kind of message addressed to itself and sets the arrival time to the desired value.

local timeout

NetSim has been implemented in Java. An actual protocol, such as SPSA, is implemented by inheriting from given generic classes that model individual node behaviour such as sending messages and joining or leaving the overlay. In order to evaluate the implemented protocol, NetSim provides a snapshot facility to perform arbitrary tasks with a global view. For SPSA, we use this to compute the overall communication cost.

NetSim's design goal of being easily transformable into a real distributed application is achieved by its modularity. Once a protocol has been implemented with NetSim, the message queue can be removed, *from simulation* and the simulated send and receive methods can be replaced by methods *to deployment* ods that actually perform these operations in the real network. Local events can be handled by timer-triggered callbacks, but the global view snapshot facility usually cannot be implemented in a distributed environment.

The message delay used by NetSim to determine the arrival time of a message can be provided either as a full distance matrix for the network, or in form of network coordinates. The distance matrix does not need *input* to be symmetric, and NetSim can also handle missing entries. Messages sent along such links can either be simply dropped, or a 2-hop alternative route can be used. If network coordinates are provided, these can be used at each node to estimate the distance to other nodes, and e. g. find the closest neighbour. NetSim also features the generation of network coordinates inside the simulation using Vivaldi as presented in Section 4.4.3.

If scalability becomes a significant issue, NetSim can also use network coordinates instead of a distance matrix for the message delay calculation. As network coordinates introduce estimation errors, cannot *alternative RTT* contain triangle inequality violations, and may also change over time, *sources* this will reduce the accuracy of the simulation. However, the modularity of NetSim also allows the use of other, more complex network generator techniques that cover more properties of real world networks.

Related Work

NetSim is a simulator especially designed to fit our needs when simulating topology optimization protocols, such as the SPSA. However, there are many alternatives for simulating P2P overlays or other network traffic. This section describes some of them.

Stephen Naicken *et al.* presented two surveys covering a wide range of P2P simulators in [127, 128]. All listed simulators are free software, *survey* with different kinds of license policies. Most of the P2P overlay simulators are implemented in Java, some used C or C++. Naicken *et al.* also introduced general criteria for the evaluation and differentiation of P2P simulators and found that none of the simulators fits all requirements.

So researchers often still conclude that no existing simulator properly fits their specific purpose.

NS-2

A powerful network traffic simulator is *NS-2* [77]. NS-2 is mainly designed for network layer simulations. It uses a very detailed model for simulating the network, including aspects of the link layer and the physical layer, which limits its scalability. Simulating these layers is not very important in P2P research, as this research is mostly concerned with the application layer. Also when simulating topology optimization protocols, we are only interested in the message delay.

OverSim

When compared to other event-based P2P simulators that concentrate on the actual P2P overlay rather than the layers below, NetSim proves to be different. *OverSim* is a P2P simulator with a focus on scalability [12]. One design goal for OverSim was an easy code reuse for practical application. However, the modelling of node-to-node delays proposed by OverSim is too simplistic: the nodes are placed in a two-dimensional Euclidean space. A more sophisticated modelling of the delays is proposed by Gerald Kunzmann *et al.* in [95]. Here, delays between hosts were modelled using network coordinates to circumvent the need for node-to-node delay matrices. Both approaches permit the simulation of large numbers of nodes but prevent the estimation of embedding errors due to the lack of actual measurements. Also, triangle inequality violations cannot be modelled in these simulators.

modelling transmission times

P2PSim
Overlay Weaver

Simulators limited to structured overlays include *P2PSim* [99] and *Overlay Weaver* [161]. While P2PSim is an event-based simulator and simulates message delay, Overlay Weaver is an emulator using four threads for each node, but neglecting the message delay. The advantage of this approach is that the emulation can easily be distributed, incorporating real networks in the emulation.

PlanetSim, PeerSim

Like NetSim, *PlanetSim* [56] and *PeerSim* are implemented in Java and support both structured and unstructured overlays. They still cannot be efficiently used for simulating topology optimization protocols, as PlanetSim offers no support for gathering statistics, while the support for latencies in PeerSim is not documented [127].

4.4.5 *Experiments*

Using the simulation framework NetSim, we have performed event-oriented simulations to evaluate the topologies constructed by SPSA. SPSA operates in the simulated network, appoints super-peers and assigns edge peers based on the perceived network delay.

We again used the PlanetLab instances as described in Section 3.3.3. As these networks only contain up to 419 nodes, we include two additional distance matrices. One matrix is from the King project [60]. *instances* King determines network latencies between arbitrary nodes by issuing recursive domain name lookup queries. In a way, it only measures the latency between the responsible domain name servers. However, the project has been successfully applied and a distance matrix for 1740 nodes is provided. We will refer to this matrix as the King matrix. Another matrix was used in the Meridian project [186]. In the paper, a distance matrix for 2500 nodes is presented. We will refer to this matrix as the Meridian matrix. Meridian actually does not provide techniques to create distance matrices, but rather provides a network location service that does not need network coordinates. The Meridian matrix was produced using the King project.

Input data was flawed. Similar to the PlanetLab instances, King and Meridian also contained missing entries and triangle inequality violations. In our experiments, we have dealt with the issue of missing distance information by computing a 2-hop alternative route. All con- *2-hop* sidered distance matrices contained flaws of that kind, which were suc- *replacement* cessfully treated. An ideal distance measurement mechanism would return the minimum or median of a sufficiently large sample of measurements. Other entries in these distance matrices were heavily distorted, e. g. the longest RTT in the Meridian matrix measured over 90 s. These entries were left unchanged; messages sent along such links introduce additional concurrency.

The simulation covered major aspects of self-organizing P2P behaviour as edge peer connections to super-peers were established, rejected, or dissolved. Each node was restricted to a local view of the network, which consisted of its connection to the super-peer and a possibly outdated super-peer list. All nodes joined the network at the same *simulation* time. One node was randomly chosen as the first super-peer. The reor- *setup* ganization cycle of SPSA was called every $\gamma = 4000 \pm 20$ ms. This value was chosen as a balance between low message complexity and swift reaction to topology changes. The random element was introduced to remove unwanted synchronicity between the nodes. In the reorganization cycle, an edge peer would try to connect to a closer super-peer, and super-peers would perform one of SPSA's reorganization steps from Section 4.4.2.

Each simulation of SPSA lasted 2500 s simulated time. NetSim used time units of 1 ms, as most of the RTT data was provided using this accuracy. NetSim's global view snapshot facility was used to evaluate the average communication cost d_{avg} every second. Each simulation was repeated 30 times and average values are used for the following discussion.

expectations

We expect simulation results to be influenced by two error sources in particular. The first source is the error introduced by network coordinates. The used network coordinates methods cannot cover triangle inequality violations, and the embedding error is further increased when the heuristic approaches for computing these coordinates fail to find the optimal solution [193]. The second error source, introduced by the distributed algorithm, is the gap between the optimal solution for the SPSP and the proposed super-peer selection.

Network Coordinates

We used Vivaldi, GNP* and GeoLite City to generate network coordinates for the networks. The coordinates created by Vivaldi and GNP* were 4-dimensional Euclidean coordinates. The geographical coordinates are 2-dimensional spherical coordinates.

Vivaldi

Vivaldi was implemented in NetSim. Each peer picked random initial coordinates from the interval [0 ms, 500 ms], using a uniform distribution. In the used 4-dimensional space, the initial distance between any two nodes did not exceed 1000 ms. During the simulation, peers were allowed to move in this embedding space according to the Vivaldi rules. They sent requests for distance estimation to other peers every 2000 ms. These requests were routed through the super-peer network to random peers, which then initiated a three-way handshake to determine the actual RTT on the direct link to the original peer and adapted the Vivaldi coordinates of both peers accordingly. Figure 35 depicts an example where peer p_1 issues a ping request to its super-peer s_1. The request is relayed to a random super-peer s_2, which again randomly decides to forward the request to one of its edge peers p_2. When the request arrives at p_2, it responds to the request, contacts p_1 and initiates the Vivaldi spring calculations.

GNP*

With GNP*, static coordinates were computed at the experiment's start with no subsequent update. We abstracted from the actual node coordinates exchange and assumed that super-peers knew the coordinates of all remaining super-peers as well as their attached edge peers, while edge peers received coordinates of other super-peers from their

Figure 35: Picking a remote peer for Vivaldi's RTT estimation. The request for a RTT measurement is forwarded to a random peer through the super-peer. The random peer and the initiating peer exchange a three-way handshake to measure the RTT and exchange their network coordinates.

assigned super-peer. In practice, this is achieved by having every node piggyback its coordinates on all outbound messages.

We used GeoLite City [106] to provide geographical coordinates. Un-   *Geo*
fortunately, no IP addresses were provided in the King and Meridian matrices, so we were unable to generate meaningful geographical coordinates in these two cases.

Lower Bounds

In order to establish a basis for comparisons, we calculated different lower bounds and also used the best known solutions found by the ELS from Section 4.3. These bounds are shown in Table 20. In this table, we use average communication times as they provide more intuitive values.

The APSP lower bound gives the cost when all communication in the network is routed over shortest paths. This is a basic lower bound for   *APSP*
all types of topologies, however, it is often far below an optimal solution for the SPSP.

The lower bound LB1 as defined in [132] is based on a relaxed MIP formulation for the USApHMP. It represents more properties of the SPSP,   *LB1*
but still produces very weak bounds when used in networks with tri-angle inequality violations. Since all considered networks have a high level of triangle inequality violations, the LB1 bound is only marginally higher than the APSP bound. The Meridian network was already too large for the LB1 calculations. CPlex [76] used up all 12 GB of memory before it could produce any values.

Network	Size	APSP	LB1	Direct	Best	Excess	Rand
01-2005	127	153 ms	156 ms	251 ms	182 ms	18.9 %	3.73
02-2005	321	154 ms	158 ms	204 ms	181 ms	17.3 %	3.09
03-2005	324	164 ms	168 ms	222 ms	198 ms	20.5 %	3.33
04-2005	70	137 ms	143 ms	165 ms	153 ms	11.6 %	2.67 ·
05-2005	374	124 ms	128 ms	209 ms	184 ms	48.3 %	3.07
06-2005	365	137 ms	141 ms	225 ms	166 ms	21.0 %	3.32
07-2005	380	173 ms	176 ms	270 ms	215 ms	24.6 %	3.39
08-2005	402	171 ms	175 ms	247 ms	192 ms	12.0 %	3.49
09-2005	419	132 ms	135 ms	238 ms	188 ms	42.1 %	3.27
10-2005	414	166 ms	169 ms	271 ms	192 ms	16.1 %	4.21
11-2005	407	143 ms	146 ms	211 ms	169 ms	17.6 %	3.27
12-2005	414	119 ms	122 ms	388 ms	167 ms	39.9 %	7.26
King	1740	99 ms	104 ms	184 ms	176 ms	77.4 %	3.11
Meridian	2500	13 ms	—	80 ms	47 ms	246.7 %	5.22

Table 20: Lower bounds for the SPSA. For each of the considered networks, the size and lower bounds based on All-Pairs-Shortest-Paths, a relaxed MIP formulation LB1, and direct connection between all nodes are shown. Also shown is the best known solution found by the ELS, and the excess of this solution above to the APSP lower bound. A comparison to unoptimized random configurations is given in column Rand as the quotient of those configurations' costs to the best known solution.

Direct

Using direct connections for all communication results in high communication costs. Comparing this value to the APSP bound gives an insight in the amount of triangle inequality violations. In the Meridian network or in 12-2005, the mismatch between APSP and direct connections exceeds a factor of 3, indicating that network coordinates will most likely fail to cover relevant properties of these networks.

Best & Excess

The best known solution found by the ELS is shown in column Best. Its excess over the APSP bound shows that realistic super-peer topologies have a cost of about 10 % to 50 % higher than the APSP bound. For King and Meridian, the cost is higher, as the triangle inequality violations are larger in these networks. However, for all considered networks, the best known super-peer topology is better than the direct connection topology, again because of the high amount of triangle inequality violations in these networks.

The comparison of the best known super-peer topologies to unoptimized super-peer topologies shows the benefit of the optimization. Column Rand gives the quotient of the cost of such unoptimized topologies to the optimized topologies. These values are calculated by generating 30 random topologies where super-peers were picked at random and edge peers were assigned to random super-peers, simulating an unsophisticated super-peer topology construction oblivious of the RTTs in the network or any notion of closeness. The values suggest that the communication cost in the considered networks can be reduced by a factor of about 3, and in some cases up to 7. *Rand*

Comparing the super-peer topologies created by SPSA to direct connections or the APSP topology reveals further advantages. Using only shortest paths for every communication gives the best possible communication cost, but also brings the necessity of determining all these routes and storing them at every node. When using direct connections, the communication cost increases, as well as the number of links each node has to maintain. In the super-peer topologies edge peers need only one link to their super-peer, and even the super-peers maintain fewer connections than nodes in a direct connection topology. Still, every communication can be achieved using only three hops without excessively increasing the communication cost. *comparison*

Results

The results of the SPSA experiments are shown in two tables. While Table 21 shows the excess of SPSA's topologies over the best known super-peer topology, Table 22 shows the gain that can be achieved by SPSA compared to unoptimized random topologies.

Although each simulation lasted 2500 s simulated time, SPSA already converged long before the simulation has finished. Different network coordinate mechanisms produced different convergence times. As all simulations converged before 900 s simulated time, we use the results at that point for our comparisons.

In the first set of experiments, we used a perfect embedding, i. e. each node knew its distances to every other node without any embedding error. This removed the error introduced by network coordinates, leaving only the error of SPSA as the total error. In practice, this means meas- *perfect embedding*

Network	Perfect	GNP*	Vivaldi	Geo
01-2005	9.6 %	55.6 %	50.5 %	76.3 %
02-2005	5.4 %	48.9 %	31.5 %	57.9 %
03-2005	6.4 %	52.7 %	29.2 %	71.0 %
04-2005	4.5 %	34.8 %	20.3 %	114.1 %
05-2005	8.5 %	47.8 %	31.0 %	62.4 %
06-2005	7.8 %	70.3 %	48.9 %	103.2 %
07-2005	11.6 %	42.0 %	45.5 %	57.5 %
08-2005	5.9 %	44.4 %	46.2 %	73.3 %
09-2005	5.9 %	49.7 %	49.1 %	74.6 %
10-2005	8.4 %	69.0 %	64.0 %	188.6 %
11-2005	5.7 %	60.6 %	45.6 %	91.0 %
12-2005	16.5 %	158.8 %	136.6 %	211.5 %
King	3.6 %	38.1 %	37.9 %	—
Meridian	4.6 %	174.1 %	137.4 %	—
average:	7.5 %	67.6 %	55.2 %	98.4 %

Table 21: Excess over the best known solution for SPSA using different network coordinates. The values are taken 900 s after the start of the simulation, such that both the dynamic network coordinates and the SPSA had sufficient time to converge.

uring the weights of all $\mathcal{O}(n^2)$ edges, or sending $\mathcal{O}(\sqrt{n^3})$ messages in each round, respectively. Thus, the results obtained using perfect embedding are not representative for a real-world application, but they give information on the quality of SPSA's work compared to the best known solutions.

Column Perfect in Table 21 and Table 22 shows these results. SPSA performed outstandingly well with a perfect embedding. The excess is seldom higher than 10 %, even though SPSA uses only local knowledge and the heuristic of connecting edge peers to their closest super-peer neglects the effects on the intra-core links. As a side note, all topologies created with perfect knowledge are superior to the direct connection topology.

The comparison to unoptimized random topologies provides new results. In these experiments, we produced 30 random topologies for

Network	Perfect	GNP*	Vivaldi	Geo
01-2005	42.3 %	10.4 %	26.3 %	6.8 %
02-2005	21.4 %	7.6 %	39.0 %	21.9 %
03-2005	20.8 %	6.8 %	41.4 %	15.6 %
04-2005	16.6 %	3.7 %	30.1 %	1.1 %
05-2005	19.5 %	19.0 %	48.5 %	22.8 %
06-2005	23.3 %	13.4 %	59.1 %	22.0 %
07-2005	16.2 %	20.9 %	32.2 %	54.8 %
08-2005	23.9 %	23.6 %	51.7 %	55.1 %
09-2005	21.8 %	16.0 %	48.7 %	55.9 %
10-2005	26.1 %	13.1 %	55.2 %	−7.4 %
11-2005	25.8 %	14.3 %	61.6 %	17.6 %
12-2005	25.9 %	9.9 %	53.7 %	93.6 %
King	24.5 %	21.4 %	54.6 %	—
Meridian	79.3 %	22.4 %	54.5 %	—
average:	27.7 %	14.5 %	46.9 %	30.0 %

Table 22: Gain of topologies optimized by SPSA against random super-peer topologies using different network coordinates. In these random topologies, super-peers were selected at random, but edge peers were connected to their closest super-peer based on the RTT estimation given by the network coordinates.

each network by picking super-peers at random, but assigned the remaining edge peers to their closest super-peer. These random topologies are superior to previously used random topologies (e. g. Table 20), as the edge peer connections are already improved. Table 22 compares SPSA results against the average cost from these random topologies and gives an insight in how much can be gained by picking the best set of super-peers. With perfect embedding, SPSA can reduce the average communication cost by between 16 % and 79 %.

Regarding network coordinates, best results can be achieved using either Vivaldi or GNP*. Using GNP* has an advantage over Vivaldi, as it already provides good coordinates at the start of the simulation. SPSA can already rely on these estimations and start improving the super-peer topology as soon as the network is formed. Thus, SPSA re- *GNP** quired little time to find good topologies. After only 40 s to 60 s, equivalent to ten to fifteen rounds of SPSA, the total communication cost had arrived at the values shown in Table 21. The resulting communication cost is higher than with a perfect embedding. Values between 40 % and 60 % above the best known solutions are usual. For networks with a high level of triangle inequality violations, such as 12-2005 or Meridian, the excess is even higher than 150 %.

Comparing the SPSA results against random topologies in Table 22 shows that SPSA can still find improvements of up to 23 %. In these random topologies, the assignment of edge peers to the closest super-peer was also based on the GNP* coordinates. In essence, using GNP* helps reducing the measurement effort of SPSA, and allowed it to quickly reach its optimization goal, creating topologies that are better than unoptimized networks.

Vivaldi

The best network coordinates in our experiments were produced by Vivaldi. In this setting, nodes update their coordinates and estimate distances to remote nodes while SPSA is working. Thus, in the beginning of the simulation SPSA establishes topologies that largely resemble random topologies. Also, coordinates produced by Vivaldi do not fully converge to a stable state, although Vivaldi's oscillation reduction [33] alleviates the lack of convergence. Due to these initial fluctuations, SPSA takes about five minutes simulated time to converge. As Vivaldi is able to adapt to changes in the underlying network, we consider it a useful alternative to enable the generation of network coordinates.

Once Vivaldi has created good coordinates, SPSA produces super-peer topologies that are superior to GNP*. The excess above the best known topologies is reduced to 30 % to 50 %. Again the networks King and 12-2005 produce larger results than other networks, with an excess of about 140 %.

Comparing SPSA's results using Vivaldi coordinates against random topologies shows improvements of up to 61 %. SPSA successfully reduces the communication cost, yielding substantial gains compared to unoptimized topologies. The nodes in a P2P system benefit from SPSA as their communication delays are reduced.

Geo

Worst results are produced when using geographical coordinates. With gaps of more than 100 % to the best known solutions even for the smallest networks, these topologies cannot be considered as a good alternative for P2P networks. There are mainly two reasons for this bad performance: Firstly, geographical coordinates do not incorporate in-

dividual link capacity, utilization or routing policies, as suggested in
[130]. Two peers in the same city but with different connections to the
Internet would still receive similar geographical coordinates. Secondly,
the quality of the coordinates also depends on the underlying database.
In the case of GeoLite City, we found that all Chinese PlanetLab nodes
were mapped to a single location in Beijing. An extreme example on
how the quality of network coordinates influences SPSA can be seen
in Table 22, where 10-2005 produced even worse results than random
topologies.

A comparison between the convergence rates for different network *comparison*
coordinates is shown in Figure 36, where the average communication
cost of the constructed topology is shown over time during an SPSA
run for the 07-2005 network. Using perfect embedding or static GNP*
coordinates, SPSA quickly finds a suitable topology. Convergence is
reached in under a minute. As Vivaldi needs time to establish good
coordinates, SPSA's convergence is slower. However, after about five
minutes there is no further significant change in the communication
cost. When using geographical coordinates, SPSA converges almost as
fast as with perfect embedding or GNP* coordinates. However, due
to the high embedding error and the fact that a large number of nodes
received the same coordinates, the load on the super-peers is not evenly
distributed. New super-peers created by SPSA soon downgrade back to
edge peer level because they only receive a very small number of edge
peers. These fluctuations can be seen clearly in the plot.

4.4.6 *Discussion*

This section presented SPSA, a fully distributed algorithm for the self-
organized construction and maintenance of super-peer topologies with
minimum communication cost. SPSA heuristically solves the problem *SPSA*
of finding a set of super-peers and the assignments of the remaining
edge peers to the super-peers, requiring only local view. By means
of simulation, we showed that SPSA converges quickly while setting
up topologies substantially better than random solutions. With GNP*,
SPSA arrived at a stable state in less than a minute, with Vivaldi in less
than five minutes of simulated time. Moreover, SPSA can automatic-
ally repair overlay partitions and adapt the number of super-peers to
the network size. SPSA and some of the results were already published
together with Matthias Priebe in [116].

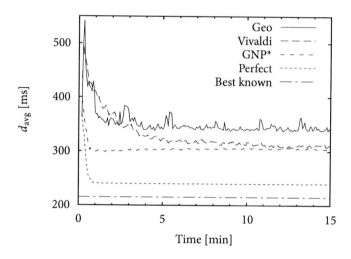

Figure 36: Comparing SPSA with different embeddings. The average communication cost is plotted over time as SPSA optimizes the topology in 07-2005 using either geographical coordinates, Vivaldi, GNP*, or a perfect embedding. Also shown is the cost of the best known solution.

The results of SPSA could be improved by allowing the nodes to measure the RTT on each link when making a decision. However, this would not scale well with the network size, the measurement traffic would counteract the goal of SPSA to reduce the load on the underlying network. Network coordinates allow SPSA to be used in large-scale settings, as they provide RTT estimations using only few RTT measurements. We described both static and dynamic mechanisms for generating these coordinates. As these network coordinates can only give estimations of the RTT on a link, the communication cost of the topologies generated by SPSA increases, depending on the quality of the network coordinates.

network coordinates

For the simulation of SPSA, we developed NetSim, a lightweight, event-based simulation framework for P2P overlay construction and maintenance that gives special attention to accurate message delay, network coordinates, and monitoring of the global communication cost. NetSim provides the opportunity to re-use the code for the simulated

simulation

protocol, as it may be easily transformed into a real distributed application with only minor modifications to the surrounding framework. NetSim was also already published together with Matthias Priebe in [115].

A disadvantage of the topologies produced by SPSA lies in the fully meshed core. In SPSA, each super-peer is expected to maintain connections to all other super-peers. This allows to send messages between edge peers using only three hops. However, if the P2P system grows, the load on the super-peer grows with $\mathcal{O}(\sqrt{n})$. This load cannot be reduced in these topologies. Assuming an equal distribution of the edge peers, the load on the super-peers is already optimal. A further reduction of this load can only be achieved by changing the topology.

fully meshed core

4.5 DISTRIBUTED ALGORITHM: CHORDSPSA

In this section, a second distributed algorithm for creating super-peer-enhanced topologies is presented. The topologies created by SPSA used a fully meshed core. This could put a high load on the super-peers especially in large networks, as they are required to maintain $\mathcal{O}(\sqrt{n})$ connections to other super-peers and edge peers. ChordSPSA replaces this core topology with a Chord ring [167], which reduces the number of intra-core links. It also increases the number of super-peers compared to SPSA, such that each super-peer now serves a lower number of edge peers.

replace the core

An example of a the type of topology constructed by ChordSPSA is shown in Figure 37. Edge peers are assigned to super-peers, which are organized in a Chord ring. Only super-peers receive a Chord ID, which is displayed in the super-peer nodes. In this topology, all communication is routed from an edge peer to its super-peer, over the Chord fingers to the destination super-peer, and then directly to the associated edge peer.

Just like SPSA, ChordSPSA is a self-organizing super-peer topology construction and maintenance algorithm that aims at low end-to-end message delay. By combining a Chord core with the super-peer hierarchy, ChordSPSA enables an efficient broadcast scheme and a reduced average message hop count compared to pure Chord. ChordSPSA also uses network coordinates to improve the assignment of Chord IDs and modifies Chord's message routing scheme to reduce end-to-end message routing delay inside the Chord ring.

improve Chord

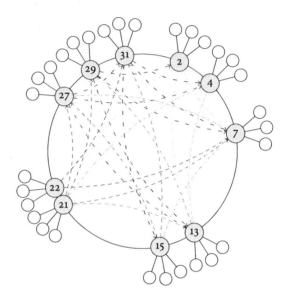

Figure 37: A Chord-enhanced super-peer overlay. Each super-peer is drawn at its position in the Chord ring, which consists of 2^5 IDs. Each super-peer maintains up to five Chord fingers to other super-peers. Edge peers are connected to one super-peer each.

support desktop grids

ChordSPSA was designed to be used especially in desktop grids. It addresses common requirements of desktop grids by providing a scalable topology with low end-to-end message delay that allows an efficient broadcast for resource discovery. There is no single point of failure in the topologies constructed by ChordSPSA. Locality is observed by clustering edge peers around the super-peers. Super-peers also allow peers to join the overlay that cannot be contacted from other peers due to firewalls. If these peers can only establish out-bound connections, they can still be contacted from other peers by having their super-peer forward in-bound messages.

This section is organized as follows. In Section 4.5.1, we provide an overview of related work. Our distributed algorithm ChordSPSA is shown in Section 4.5.2. In Section 4.5.3, we propose Chord modifications to reduce the delay inside the Chord ring. In Section 4.5.4, we present results of simulations, which show the benefit of ChordSPSA. The section on ChordSPSA is concluded in Section 4.5.5.

4.5.1 Related Work

Desktop grids have been proposed to make use of the idle computational resources provided by standard desktop computers. An empirical study showing the amount of idle resources is presented by Derrick Kondo *et al.* in [90]. They evaluated the availability of over 220 hosts over time, and calculated expected task failure and work rates. They also evaluated the cluster equivalence ratio of the desktop grid, i. e. the number of dedicated cluster nodes the desktop grid could replace. They concluded that their set of 220 hosts was at least equivalent to a 160-host dedicated cluster. *desktop grids*

An early form of distributing computational tasks to idle desktop computers is volunteer computing. These systems usually serve a single task, and participants volunteer to contribute because of some personal or general interest in this task. A popular technique behind a large number of such public-resource computing projects is BOINC, the Berkeley Open Infrastructure for Network Computing [5]. In BOINC, an asymmetric relationship between the projects and the participants is established. Each project maintains its own servers, and volunteers can contribute their idle computation resources by running a dedicated client. *BOINC*

In [6], Nazareno Andrade *et al.* present OurGrid, a P2P desktop grid where each participant can submit jobs. The authors concentrate on incentive mechanisms and design a *Network of Favors*. Using these mechanisms, each peer assigns scores to other peers based on former interactions, and uses these scores to decide whether or not to donate its own computation resources. *OurGrid*

Arjav J. Chakravarti *et al.* present OrganicGrid in [23], a self-organizing P2P desktop grid. In this grid, a tree topology is used, where the initiator of a task is the root. The tree topology is periodically reshaped to move faster nodes closer to the root. As with TreeOpt (Section 3.4.3), the creation of cycles has to be avoided and unintentionally created cycles have to be dissolved. *OrganicGrid*

All these studies showed that the cumulative potential of a desktop grids grows with the number of resources connected to it. Therefore, desktop grids benefit from the capability to integrate a heterogeneous range of computers in a dynamic and decentralized environment. Scalable P2P networks can be used to build a desktop grid structure [74]. The scalability of P2P networks can further be improved by the concepts of super-peers [80, 191, 194] and structured P2P overlays [61, 104, 167].

Chord

ChordSPSA combines super-peer topologies with flat P2P structures by introducing a super-peer hierarchy on top of Chord. Chord was introduced by Ion Stoica *et al.* in [167] to provide a distributed hash table storing values at well defined locations in the P2P overlay. ChordSPSA does not use this functionality, only the finger-enhanced ring structure is used. Chord establishes this self-organizing ring-structured P2P overlay in a scalable, decentralized fashion. Every Chord node assigns itself an ID drawn uniformly from a sufficiently large identifier space when joining the Chord overlay. This ID defines the position of this node in the ring. In addition to the connections to its successor and predecessor in the ring, each node maintains further links. These links are called the *fingers* of this node. The fingers can be seen as shortcuts, covering all power-of-two distances in the Chord ID space. When forwarding a message to a given ID, the longest finger that points directly to this ID or to a position before this ID is used. Thus, messages require $1/2 \cdot \log_2 n$ hops in the average case. Chord also scales very well, as each node maintains only $\mathcal{O}(\log n)$ finger connections to other peers in a network of n nodes. The finger set is updated periodically to allow an adaptation to changes in the P2P network, such as leaving or joining nodes.

$\mathcal{O}(1)$ Chord

An unusual example of integrating super-peers in Chord is given by Alper T. Mızrak *et al.* in [121]. They use Chord in an outer ring of their topology, but also create an inner ring which consists of all super-peers. Each super-peer is responsible for a section of the Chord ring. A lookup is performed by transferring the request to the responsible super-peer, which returns the address of the responsible peer in the Chord ring. Using this topology, all lookups can be achieved with at most three hops. However, the super-peers have to maintain links to all other super-peers, a disadvantage ChordSPSA explicitly wishes to avoid.

proximity awareness

The importance of proximity awareness in P2P overlay management has been stressed by P. Krishna Gummadi *et al.* in [61]. They concentrate on various P2P geometries and find the ring geometry, as established by Chord, to be most flexible and resilient. Gummadi *et al.* also point out that proximity-related optimizations may be performed during routing and identifier assignment.

HP-Chord

An example for a proximity-aware variant of Chord is given by Feng Hong *et al.* in [70]. The authors introduce *HP-Chord*, modifying the routing component of Chord. They propose to create multiple virtual nodes for each peer. These virtual nodes are placed in the Chord ring and a proximity list is kept to identify nearby neighbours. The scalability of this approach remains questionable as HP-Chord explicitly relies on live delay measurements and does not make use of network coordinates.

4.5.2 Chord Super-Peer Selection Algorithm

ChordSPSA is based on SPSA, it uses similar rules for super-peer promotion and edge peer assignment. These rules are adapted to accommodate a Chord core. As with SPSA, every peer periodically enters a reorganization phase and tries to improve the overall topology based on its local knowledge.

Number of Super-Peers

In ChordSPSA, peers promoted to super-peer level join the Chord ring, while super-peers downgrading to edge peer level leave the ring. Hence, the set of super-peers matches the set of Chord ring members. As in SPSA, ChordSPSA uses the number of intra-core links to determine the maximal number of edge peers $|E_k| - 1$ a super-peer k should serve. This upper limit is calculated by the number of distinct Chord fingers f and a constant factor m:

$$|E_k| - 1 \leq f \cdot m \tag{4.22}$$

The number of fingers f grows logarithmically with the core size $|C|$. In a perfect Chord ring it is $f = \log_2 |C|$. The factor m is a parameter of ChordSPSA and can be used to select the ratio between super-peers and edge peers in the network. For $m = 0$, super-peers do not accept any edge peers, and the resulting topology is a pure Chord ring. For $m \geq n - 1$, a star topology with only one super-peer is created. As one aim of ChordSPSA is to reduce the load for the super-peers, small constant values should be used, e. g. $5 \geq m \geq 20$.

parameter m

The number of super-peers x in a perfect ChordSPSA topology could be calculated by numerically solving the following equation:

$$x \cdot \log_2 x \cdot m + x = n \tag{4.23}$$

However, ChordSPSA uses simple rules and does not depend on knowledge of the network size n to approximate this value.

promotion

In every reorganization phase, the super-peers compare their number of edge peers $|E_k| - 1$ to the upper limit given by $f \cdot m$. If a super-peer serves too many edge peers ($|E_k| - 1 > f \cdot m$), it promotes one of its edge peers to super-peer level. Some of the remaining edge peers will then switch to this new super-peer. Following this rule, the number of super-peers grows as long as unserved edge peers require more super-peers. Once every edge peer is managed by a super-peer, no further super-peers are created. The minimum number of super-peers is reached when every super-peer administrates the maximum number of edge peers. In ChordSPSA, the number of edge peers for each super-peer grows logarithmically with the overlay size n compared to $\mathcal{O}(\sqrt{n})$ in the SPSA.

downgrade

If a super-peer serves a very low number of edge peers, it downgrades to edge peer level. The lower bound is determined by $1/k \cdot f \cdot m$. Again, k is a parameter of ChordSPSA. If set too low ($k < 2$), the creation of a new super-peer will cause either the new or the old super-peer to be underloaded, as the remaining $f \cdot m$ edge peers are not sufficient to increase the load on both super-peers to the minimum level. In such settings, one of these super-peers would downgrade, and the remaining super-peer would again appoint another super-peer. As this scenario never converges, unless edge peers switch to other super-peers, $k = 2$ gives a lower bound.

Even with this value, convergence is unlikely. The new super-peer can only stay active if it receives half of the edge peers. The same applies to the old super-peer. So with $k = 2$, only an even split would allow both super-peers to stay active. A more realistic setting of $k = 4$ is used in the experiments in Section 4.5.4. Although this allows an uneven distribution of the edge peers, it reduces the amount of unnecessary upgrades and downgrades.

Delay Minimization

hop count

Since the super-peer structure adds two additional hops to message paths when two edge peers communicate, the choice of m determines the average hop count. In the super-peer overlay, the average hop count for communication between two edge peer is $h_{avg} = 2 + 1/2 \cdot \log_2 x$. It is smaller when one or both communication endpoints are super-peers. The average hop count in a pure Chord ring is $1/2 \cdot \log_2 n$. So, if each super-peer serves at least 15 edge peers ($n/x \geq 2^4 = 16$), the super-peer overlay already reduces the average hop count. This can be achieved by selecting an m sufficiently large such that $f \cdot m > 15$. As f grows logarithmically with the number of super-peers, small constant values for m are already sufficient to reach this goal.

For the actual communication cost, we use the following equation:

$$Z = \sum_i \sum_{j \neq i} \underbrace{d(i, s(i))}_{\text{spoke}} + \underbrace{d_{\text{Chord}}(s(i), s(j))}_{\text{Chord}} + \underbrace{d(s(j), j)}_{\text{spoke}} \quad (4.24)$$

Again, we use $s(x)$ to denote the super-peer for peer x, and set $s(x) = x$ if x is a super-peer. The distance $d(x, y)$ refers to the actual network delay between nodes x and y, and $d_{\text{Chord}}(x, y)$ refers to the distance between two Chord nodes in the Chord ring, taking into account all hops necessary for the Chord routing. As Chord uses different routes for each direction, d_{Chord} is not symmetric. The average message delay d_{avg} in a network of size n is calculated as $d_{\text{avg}} = \frac{Z}{n \cdot (n-1)}$. *communication cost*

When an overloaded super-peer k selects one of its edge peers E_k to be promoted as a new super-peer, it already performs a check to determine whether both k and the new super-peer will receive a sufficient number of the edge peers after they have switched to the closest super-peer. This is done to prevent an instant downgrade in the next reorganization phase. The super-peer can perform this check based solely on the known network coordinates of its edge peers. Out of all edge peers that fulfil this requirement, the edge peer e is chosen which results in the minimal total finger delay. *picking super-peer candidates*

ChordSPSA manages edge peers in the same way as SPSA. Edge peers connect to their closest super-peer. As edge peers can be expected to be unevenly distributed among the super-peers, some super-peers will serve fewer edge peers than their maximum load would allow. This slightly increases the number of super-peers created by ChordSPSA beyond the optimal value given in (4.23). *edge peers*

Messages

ChordSPSA uses similar messages as SPSA to construct and maintain the super-peer topology. Edge peers can query a list of super-peers using a QUERYSUPERPEER message. The super-peer list sent in the reply also contains currently known network coordinates for these super-peers to allow the edge peer to determine the nearest super-peer. A connection to this super-peer is established using a CONNECT message. Each edge peer also keeps a local super-peer taboo list to prevent re- *SPSA messages*

Chord messages

peated connection attempts to the same super-peer. Super-peers appoint new super-peers using a NEWSUPERPEER message. They also maintain the Chord ring using standard Chord messages described in [167].

When used in a desktop grid, peers may wish to find idle nodes that meet the requirements for a computational task. Unless super-peers keep a list of the configurations and status of their edge peers, this search can be performed by broadcasting the requirements and

broadcasts

waiting for replies. A broadcast in this Chord-enhanced super-peer topology can be efficiently performed by sending the broadcast to the assigned super-peer, which then follows an established broadcasting scheme, such as [7] or [110]. In these broadcasting schemes, the Chord fingers are used to distribute the message to all super-peers. In a final step, the super-peers receiving the broadcast forward it to all assigned edge peers. A sender should also be able to restrict the broadcast to reach only a limited portion of the identifier space. This is useful for desktop grids, where a job-submitting peer usually only needs a certain number of worker peers. As regular broadcasts cover the entire overlay, the wave of responses could be overwhelming.

Deterministic Gossip

super-peer list

When an edge peer wants to find its closest super-peer, it uses a super-peer list provided by its current super-peer. This list cannot be generated from the Chord fingers, as they only cover a very small sample of all super-peers. Instead, ChordSPSA maintains an explicit super-peer list at each super-peer. This list contains Chord IDs, network addresses, and network coordinates.

If a super-peer fails, its edge peers use this list to connect to other super-peers. The super-peer list also enables peers to directly deliver urgent unicast messages to the destination super-peer at the expense of setting up an additional connection. Finally, the super-peers also use this list for a delay-optimized Chord ID selection.

gossip

This super-peer list is kept up-to-date using deterministic gossipping. As peers may join and leave the overlay at any time, super-peers are created and removed, causing the ring to change continuously. Gossip intends to capture such changes and to notify other ring members. It is a method to perform anti-entropy super-peer list synchronization [42]. As gossip provides a best-effort approach for continuously keeping the list's contents current, the super-peer list is only loosely consistent.

All peers log recorded Chord ring changes in a local *update set*. The *update set* elements in this set describe an upgrade or a downgrade event for one node. They store the Chord ID of this node, its recent network coordinates, a propagation counter, and a sequence number allocated by a scheme presented in [142]. Using this scheme, new information can override old information, as entries with a higher sequence number override entries with a lower sequence number for the same peer. Odd sequence numbers mark peers that do not belong to the Chord ring, while even sequence numbers mark live Chord nodes. When a Chord node notices the failure of one of its finger peers, it increments the corresponding sequence number and propagates this information using gossipping.

Each super-peer in ChordSPSA periodically transmits its super-peer list updates to one of its finger peers, iterating through its finger set in a round-robin way, and receives a reply containing the update set of the corresponding super-peer. Additionally, the super-peer forwards *round-robin* received updates to its edge peers. This deterministic approach to gossipping allows to give an upper bound on the propagation time. In a Chord ring with x members, the longest path contains $\log_2 x$ hops. As each intermediate node follows the round-robin procedure, in each step, one of the $\log_2 x$ fingers of this node is chosen. The correct finger for the longest path will be chosen after at most $\log_2 x$ steps. Thus, after at most $(\log_2 x)^2$ steps, the update is propagated to all nodes.

With this gossip scheme, some node will receive the same update multiple times. This introduces necessary redundancy to cope with churn. While a simple broadcast tree with $x - 1$ edges would suffice, the departure or failure of a single node would break this tree, leaving some nodes uninformed. The gossip scheme uses in ChordSPSA is more robust, but still efficient.

4.5.3 *Delay Minimization in Chord*

The routing scheme in Chord was designed to use a minimal number of hops [167]. Messages are sent along the finger that covers the largest Chord ID distance towards the target. In a stable Chord ring with x nodes, all messages can be delivered in $\mathcal{O}(\log x)$ hops. However, as fingers are chosen by ID only – without considering properties of the link such as its bandwidth, the RTT or possible congestion – Chord's routing scheme does not provide an optimal routing.

While introducing a super-peer hierarchy already improves the average hop count and the average delay by reducing the size of the Chord ring and providing locality by connecting edge peers to their closest super-peer, ChordSPSA also aims at improving the message delay inside the Chord ring.

As pointed out in [61], there is a tradeoff between reducing the average message delay and reducing the average hop count. In the context of desktop grids, we find latency to be more important. Besides the benefit for desktop grids, low message delays also improve the self-stabilization properties of Chord as information about changes in the ring is propagated faster.

ChordSPSA uses two Chord-related optimizations. Following the terminology set up by P. Krishna Gummadi *et al.* in [61], these methods are called Proximity Route Selection (PRS) and Proximity Identifier Selection (PIS). Gummadi *et al.* evaluated the effects of PRS schemes in Chord and other structured P2P overlays and showed that they are effective in reducing the average message delay. We have designed PRS and PIS mechanisms for the Chord ring in ChordSPSA. These mechanisms can be used together or separately, they can also be used without the super-peer structure of ChordSPSA. We will show in Section 4.5.4 that the largest gains are achieved if PRS and PIS are used in combination with the proposed super-peer overlay.

Finger Use in Routing

The routing in Chord aims at minimizing the number of hops for each message. Each forwarding node selects the finger that covers the largest distance in the Chord ID space. As ChordSPSA aims at minimizing the message delay, we propose the following PRS scheme: Out of all suitable fingers, the finger that provides the highest 'speed' is chosen, i.e. the finger that covers the largest distance in the Chord ID space per unit of delay. As each super-peer knows the network coordinates of the remaining super-peers, this speed value can be calculated quickly.

PRS

Denoting the set of suitable fingers at super-peer x as F_x, the proposed PRS scheme follows this formula to select the best finger y^*:

$$y^* = \arg\max_{y \in F_x} \frac{d_{ID}(x, y)}{d_{NC}(x, y)} \tag{4.25}$$

Here, d_{ID} gives the distance in the Chord ID space, and d_{NC} gives the message delay as estimated using network coordinates.

In accordance with the Chord routing scheme, no finger that points to a location in the Chord ring behind the target is chosen. PRS may increase the number of hops for this message, but it strives to reduce the average end-to-end message delay. Due to violations of the triangle inequality, messages routed via multiple latency-optimized hops may even arrive earlier than messages sent via a direct but slow link. The proposed scheme is still a best effort scheme, as it does not take the remaining path of the message into account. In the worst case, a message may be transferred using a high-speed finger to a node that only has outgoing finger connections with a high delay. In this case, the message delay may actually be increased.

Chord ID Selection

In Chord, each peer is located in the ring by its ID. This ID is chosen randomly from a sufficiently large identifier space, [167] proposes 160-bit keys, such that the probability of assigning the same ID twice are negligible. As the peer's ID determines the fingers for this peer, the ID has influence on the message delay for all outgoing messages from this peer.

As ChordSPSA aims at minimizing the average message delay, we propose the following scheme to introduce proximity awareness when selecting Chord IDs. The newly created super-peer that wishes to join *PIS* the Chord ring already has a copy of the super-peer list, showing network coordinates and Chord IDs of all super-peers. This information is used in ChordSPSA to find the optimal position, without the need of sending further messages such as RTT measurements. In our PIS scheme, we propose to use an ID that minimizes the sum of delays *minimize finger* for all outbound finger connections. When the optimal ID is calcu- *length* lated, the super-peer joins the Chord ring following the standard Chord rules.

Due to the vast size of the ID space, considering all non-occupied positions is impracticable. Therefore, ChordSPSA limits the choice of possible IDs to all positions located halfway between two successive peers in the Chord ring. The first super-peer picks a random ID. The *middle position* second super-peer is placed directly opposite to the first peer. Every following peer is presented with a Chord ring with p occupied positions.

It checks all p middle positions and chooses the ID that minimizes its total finger delay.

As the super-peer list may contain outdated information, two peers concurrently joining the ring could choose the same ID, causing a collision. The probability of a collision is higher than in normal Chord, as the proposed PIS scheme provides very few choices for new Chord IDs. If a joining super-peer picks an already occupied ID, the ring members reject this super-peer and provide it with an up-to-date super-peer list. Each super-peer also checks the super-peer list for other super-peers that use its own Chord ID. If there are such entries, the super-peer can decide to change its ID or asks the conflicting super-peers to change their ID.

Using the PIS and PRS scheme in ChordSPSA may cause the message routing load to be unevenly distributed among the super-peers. However, ChordSPSA gives overloaded super-peers the means to promote one of their edge peers to super-peer level, thus distributing excessive load. We find the remaining uneven load distribution acceptable, considering that the overall message delay is still reduced.

4.5.4 *Experiments*

ChordSPSA, the distributed topology construction and maintenance algorithm presented in this section, was implemented in the simulation framework NetSim. As ChordSPSA replaces the fully meshed core of the super-peer topology built by SPSA with a Chord ring and uses different thresholds for super-peer promotion and downgrade, the solutions for the SPSP from the previous sections cannot serve as a base for comparison: The Chord routing increases the hop count even when the proposed Chord improvements are applied, so we expect larger values for the average message delay d_{avg} than with a fully meshed core. Instead, we compare the average message delay in the created topologies to the direct connection topology. Again, all experiments were repeated 30 times and average values are used in the discussion.

Setup

simulation   Again, each experiment lasted 2500 s of simulated time. The simulation covered major aspects of self-organizing behaviour as edge peer connections to super-peers and Chord ring links were established, rejected, or dissolved. The peers were restricted to a local view of the network, keeping a possibly outdated super-peer list. As all nodes joined the network at the same time, we chose one node at random to act as the first super-peer. The periodical reorganization phase for ChordSPSA was called every $\gamma = 4000 \pm 20$ ms, the random element was added to remove unwanted synchronicity. A gossip cycle was triggered in every reorganization phase, to keep the super-peer list up-to-date. We again used Vivaldi to create and maintain network coordinates, as Vivaldi gave the best results in SPSA. The initial 4-dimensional coordinates were chosen from the interval $[0\,\text{ms}, 500\,\text{ms}]$, using a uniform distribution. We experimented with different values for the maximum load factor m for the super-peers, but fixed the lower bound for super-peer downgrades to $k = 4$. Regarding Chord, we conducted experiments with and without the proposed Chord improvement, and chose an identifier space containing 2^{20} elements.

reorganization times

Vivaldi

super-peer load

Chord

The experiments were based on the PlanetLab delay matrices, which were already used in the SPSA experiments in Section 4.4.5. We again replaced missing entries in these delay matrices by computing a 2-hop alternative route.

instances

Results

The effect of the proposed Chord improvements as described in Section 4.5.3 can be shown when promoting all peers to super-peer level. This way, all peers are connected in the Chord ring, and the additional hops introduced by the super-peer hierarchy are removed. ChordSPSA can be used to create such topologies by setting the maximum load factor to $m = 0$. In essence, each super-peer that serves an edge peer will consider itself overloaded and promote this edge peer to super-peer level. As ChordSPSA uses reorganization cycles to create the topology, the promotion of peers takes some time, during which Vivaldi can establish good network coordinates for all peers.

Chord enhancements

The results for these experiments are shown in Table 23. The table shows the average message delay in the created Chord ring after 900 s simulated time. By this time, all peers were promoted to super-peer level, the network coordinates have converged to good values, and the Chord fingers are established. The original version of Chord without improvements is simply labelled Chord, and versions using the proposed improvements are labelled PIS, PRS, and PIS+PRS. The values show that PIS alone does not improve the average message delay. In some cases, the delay even increases. In these setups, the Chord rout-

PIS

ing does not make use of the improved finger delays of the PIS scheme. Chord routes messages along the finger that covers the largest distance in the Chord ID space, one half of the messages are routed over the longest finger, one quarter of the messages over the next finger, and fewer messages for every other finger. However, our PIS scheme places the same weight on all fingers when improving the finger delay. So in some cases, PIS improves fingers that are not frequently used in the

PRS Chord routing. The effects of PRS are better. The average delay is reduced by 23 % to 38 %. Using improved Chord IDs helps to further improve the routing. In the experiments, the average message delay was

PIS+PRS reduced by 27 % to 52 % when PIS and PRS were combined. In these setups, the improved routing makes use of the improved finger delays. Also, as our PIS scheme places new super-peers halfway between other nodes in the Chord ring, many of the positions the Chord fingers point to are occupied.

The progress of ChordSPSA can be seen in Figure 38. Here, the average message delay is plotted over time as ChordSPSA constructs and

progress improves the topology for the PlanetLab instance 09-2005. During the first minute of simulated time, ChordSPSA still promotes peers to super-peer level, so these values are not representative. After the initial Chord ring is build, ChordSPSA uses the network coordinates to improve the routing, and peers try to find better finger connections. Using pure Chord, the missing finger connections during the first five minutes increase the average message delay. When PIS is used, the finger connections can be established faster, as most of the finger positions are occupied. When PRS is used without PIS, the effects of missing finger connections is also reduced, as PRS chooses the best among all established fingers for routing. When PRS is combined with PIS, the average message delay converges quickly to about half the delay in pure Chord.

In the next set of experiments, we evaluated the influence of ChordSPSA's super-peer hierarchy. Here, we are interested in the effects of varying the maximum number of edge peers per distinct finger m on

varying m the average message delay d_{avg}. The results are shown in Table 24. We varied m between $m = 0$ and $m = 50$. As $m = 0$ simply creates a flat Chord ring, the respective results are labelled Chord. The results clearly show that the addition of super-peers brings further improvings compared to Chord enhanced with PIS and PRS. Also, the average message delay d_{avg} decreases with rising m. However, too large values for m result in excessive load for the super-peers in large networks. If, for

	average message delay d_{avg} [ms]			
Network	Chord	PIS	PRS	PIS+PRS
01-2005	1098	1036	844	668
02-2005	1054	1130	775	680
03-2005	1191	1271	831	705
04-2005	650	615	469	457
05-2005	1103	1257	823	730
06-2005	1151	1290	824	754
07-2005	1458	1337	901	756
08-2005	1272	1267	880	720
09-2005	1303	1304	867	713
10-2005	1480	1358	942	818
11-2005	1126	1319	846	817
12-2005	1787	1515	1154	856
average	1223	1225	846	723

Table 23: Impact of Chord improvements. The average message delay of pure Chord is compared to the proposed improvements PIS, PRS, and combined PIS+PRS. Each line shows the values measured at 900 s simulated time for one PlanetLab instance.

example, a super-peer maintains 10 distinct Chord fingers, a value of $m = 50$ results in a maximum load of 500 edge peers for this super-peer.

The progress of ChordSPSA for network 09-2005 is shown in Figure 39. The plot compares results for the Chord ring to super-peer-enhanced topologies using reasonable values for m. In both cases, we used the Chord improvements PIS and PRS. The average message delay was again reduced by a factor of about 2. *progress*

When comparing the results for ChordSPSA with SPSA, a disadvantage of the additional Chord routing step is revealed. While SPSA created topologies with lower average message delays than the direct connection topology, ChordSPSA's topologies have a higher delay than direct connections. In SPSA, the triangle inequality violations can be avoided by finding good three-hop routes. In ChordSPSA, the Chord routing requires more hops, even with PIS and PRS improvements. Also, PRS is limited to delay estimations by the network coordinates, *ChordSPSA vs. SPSA*

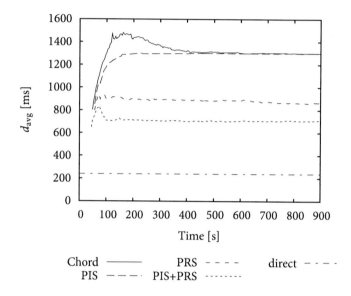

Figure 38: Impact of Chord improvements. The plots show the average message delay d_{avg} over time as ChordSPSA is working to improve the topology for the network 09-2005. Pure Chord is compared to the improvements PIS, PRS, and combined PIS+PRS. Also shown is the average message delay for a direct connection topology.

which do not model triangle inequality violations. However, the topologies created by ChordSPSA are more scalable than those created by SPSA, as the load for each super-peer is reduced. Considering that direct connection topologies are impracticable, as each peer would need to maintain connections to all other peers, the super-peer-enhanced Chord topology created by ChordSPSA is a good choice for desktop grids and other P2P systems that benefit from scalable topologies with low message delays.

4.5.5   Discussion

Chord core

This section presented ChordSPSA, which replaces the fully meshed core topology of SPSA with a Chord ring. The algorithm is again fully distributed and can be used for the self-organized construction and

	average message delay d_{avg} [ms]					
Network	$m = 0$	$m = 2$	$m = 5$	$m = 10$	$m = 20$	$m = 50$
01-2005	668	690	465	405	370	289
02-2005	680	611	491	423	363	283
03-2005	705	662	524	451	407	290
04-2005	457	285	252	228	201	200
05-2005	730	668	581	424	407	290
06-2005	754	534	556	445	353	307
07-2005	756	675	618	501	397	355
08-2005	720	588	623	457	373	323
09-2005	713	605	586	474	387	316
10-2005	818	619	613	482	406	355
11-2005	817	666	593	463	384	312
12-2005	856	856	678	588	532	513
average	723	622	548	445	382	319

Table 24: Impact of the maximum number of edge peers per distinct fingers m. The average message delay after 900 s of simulated time is shown for each PlanetLab instance. In these experiments, both Chord enhancements PRS and PIS are used.

maintenance of super-peer-enhanced P2P topologies with minimum communication cost. As the additional Chord routing step in the inner core increases the average hop count as well as the average message delay, ChordSPSA uses two Chord improvements. PIS is used to select suitable Chord IDs, and PRS is used to improve the Chord routing. We showed that the average message delay in the Chord ring can be reduced to almost half the values of the original Chord routing.   *Chord improvements*

The introduction of the super-peer hierarchy gives further reductions. ChordSPSA uses a parameter m to determine the maximum load for each super-peer. Using this value, ChordSPSA adapts the number of super-peers to the size of the P2P system. There is a tradeoff between high load for the super-peers and lower average message delay. However, we showed that even moderate settings for the parameter m allowed topologies that reduced the average message delay again by a factor of 2 compared to a flat Chord ring. Some results of ChordSPSA   *load vs. message delay*

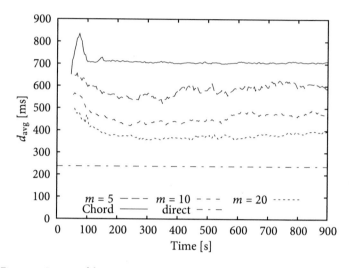

Figure 39: Impact of the maximum number of edge peers per distinct fingers m. The plots show the average message delay d_{avg} over time as Chord-SPSA is working to improve the topology. In these experiments, both Chord enhancements PRS and PIS are used.

have already been published together with Matthias Priebe and Dennis Schwerdel in [118].

network coordinates
The results of ChordSPSA heavily depend on the quality of the used network coordinates. ChordSPSA uses Vivaldi, which proved to be a good choice as Vivaldi not only produces a low embedding error, but also adapts the coordinates to changes of the RTTs in the network. However, no network coordinates mechanism can fully reduce the embedding error. One source of erroneous embeddings are triangle inequality violations, which occur frequently in the Internet. The heights of Vivaldi could be used to model triangle inequality violations, if this technique was extended to allow negative heights.

4.6   SUMMARY

Super-peer-enhanced P2P topologies that reduce the average message delay can be constructed and maintained in a self-organizing manner. This chapter has shown two distributed algorithms to achieve this goal.

These distributed algorithms differ in the kind of topology they con- *distributed* struct. While SPSA uses a fully meshed inner core, ChordSPSA con- *algorithms* nects all super-peers using a Chord ring. There are advantages to both approaches. In SPSA, every communication can be achieved with only three hops. ChordSPSA increases the hop count as well as the message delay, but reduces the load on the super-peers, which makes it more scalable.

The quality of both algorithms largely depends on good network coordinates. We evaluated three approaches. Geographical coordin- *network* ates were easy to achieve, but resulted in very bad topologies. A static *coordinates* method for creating network coordinates, GNP*, showed quite good results. Slightly better results were achieved with Vivaldi, a dynamic method for creating network coordinates. As Vivaldi also adapts the network coordinates to changes in the network, such as increased RTTs due to changing routing policies, it is the best choice for proximity aware P2P systems.

To find optimal topologies, we resorted to global view optimization algorithms. The problem SPSA solves is the Super-Peer Selection Prob- *global view* lem (SPSP). It is related to a hub location problem called Uncapacit- ated Single Assignment p-Hub Median Problem (USApHMP). Both problems are $\mathcal{NP}$-hard, so exact solvers based on Mixed Integer Pro- gramming (MIP) can only be used to solve relatively small problem instances. Therefore, we proposed an Evolutionary Local Search (ELS) heuristic that finds near-optimal solutions in short time. The ELS can be applied to the SPSP as well as the USApHMP, which allowed to com- pare the ELS to other heuristics. The ELS proved to be competitive, it found new best known solutions for well established test instances for the USApHMP.

When we started work on the SPSP, the mathematical background for hub location problems was already well studied. However, a formal *complexity* proof for the $\mathcal{NP}$-hardness of SPSP was still missing. We provided this proof by reduction from the maximum clique problem.

Even though the super-peer-enhanced topologies could be used in a large variety of P2P systems, our larger goal was to use them in a desktop grid. Desktop grids require scalability, as they profit from each *use in desktop* additional peer that can join the grid, and they need small message *grids* delays when a job initiator wants to find suitable workers quickly or when workers need to cooperate. The focus of this thesis is topology optimization, however, the work described in this chapter has already been continued by Matthias Priebe to create a desktop grid called Peer-

Grid [117, 114, 119, 144]. The ChordSPSA code was successfully transferred from the NetSim framework, methods for job submission were
included, but the topology construction remained essentially the same
as described in this section. Thus, the results in this chapter, which are
based on simulation, were also confirmed in real distributed settings.
ChordSPSA was able to handle churn, and a reputation system was added to prevent free-riding and provide incentives for peers to become
super-peers.

5

RANGE ASSIGNMENT PROBLEMS

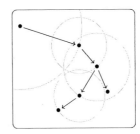

5.1 MOTIVATION

Wireless networks establish a completely different kind of optimization problems. In the previous chapters, we were mostly concerned with the average message delay in a chosen network topology. While this may still be a valid optimization goal in wireless networks, the most important objective in such networks is probably saving energy. Wireless networks can be built using standard desktop or laptop computers, however, they are also open to a vast variety of other possibilities:

Wireless sensor networks are one example. The nodes in these networks are usually small battery-powered devices, equipped with some kind of sensor and very limited computational resources. They are placed in strategically relevant positions and monitor activity in their sensor region, such as movement of cars or people, weather data, or signs of smoke or fire. This sensor data is pre-processed on the nodes, but it is not stored on the nodes, instead it is transmitted to a data sink, usually connected to a wired network and thus to a human operator. An extension to these networks also includes actuators in the wireless environment. These wireless sensor/actuator networks can then react on their own on the perceived sensor data.

wireless sensor networks

Wireless ad-hoc networks are another example. Such wireless networks can be used as an emergency system to establish a new communication infrastructure, e. g. in a disaster area, where former telephone

wireless ad-hoc networks

lines were destroyed. There are already approaches to establish such emergency networks using small robotic air vehicles [66].

energy consumption

Except for the case of mains-operated laptop computers, the nodes in a wireless network are usually battery-powered [152, 86]. They are provided with a limited amount of energy, and use this energy to power the sensor, the CPU and the transmission unit. If the energy is used up, the battery needs to be replaced, however, in some applications, this is impracticable or even impossible. Wireless sensor nodes dropped from an airplane over a forest to monitor an ongoing fire or sensors embedded in the walls of a building monitoring the structural integrity are just two examples. It is obvious that the wireless network should be optimized to use as little energy as possible.

transmissions

In this chapter, we are especially interested in reducing the energy needed for communication, as computational tasks and sensor requirements are very task-specific, but communication is needed in every type of wireless network. Communication in wireless networks can be performed either directly between two nodes, or by relaying the messages via intermediate nodes. Each node is able to adjust its transmission power based on the distance to the receiver [152, 86]. Besides saving energy, this also helps to keep the interference between different simultaneous communications low.

The necessary transmission power is usually modelled as a polynomial function of the distance to the receiver, e. g. the squared distance in the case when there are no obstacles. Thus, the total energy consumption is often lower when using intermediate nodes than when sending the message directly to the destination [152, 86]. In an example without obstacles shown in Figure 40, it is twice as expensive to send a message directly instead of sending it to a node halfway between sender and destination and having this node relay the message. In the latter case, both transmitting nodes will only use a quarter of the power of the direct connection. When there are obstacles between sender and destination, such a multi-hop connection is often the only possibility to transmit a message.

saving energy by multi-hop connections

Because of the limited battery power of each node, it is crucial to find communication topologies that minimize the energy consumption. This leads to two opposing aims when setting up wireless ad-hoc networks: The network lifetime should be high, so settings with lower transmission powers are preferred. On the other hand, the network has to stay connected, so too low transmission ranges should be avoided.

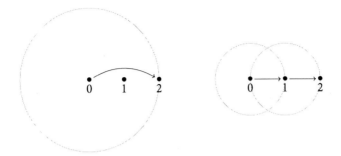

Figure 40: Energy consumption of wireless transmissions. In these graphs, the circles represent the transmission ranges of the wireless nodes, and the arcs represent established links. If the transmission power is proportional to the squared distance, the area of the circles also represents the necessary transmission power. The transmission scheme on the left uses twice as much energy as the transmission scheme on the right.

In this chapter, two problems concerning the topology construction in wireless networks are considered. Before presenting a number of global optimization heuristics and a distributed algorithm, Section 5.2 introduces the necessary mathematical background. In Section 5.3, we are searching for the broadcast tree that minimizes the total energy consumption. This problem is known as the Minimum Energy Broadcast problem. We present three global view heuristics for this problem: a Nested Partitioning algorithm in Section 5.3.2, a hybrid algorithm using Nested Partitioning and Max-Min Ant System in Section 5.3.3, and an Evolutionary Local Search in Section 5.3.4. In Section 5.4 and Section 5.5, we are searching for the wireless topology that minimizes the total energy consumption but still gives a symmetrically connected network. This problem is known as the Minimum Power Symmetric Connectivity Problem. We present a global view heuristic based on Iterated Local Search in Section 5.4.2 and a distributed algorithm in Section 5.5.2. The algorithms and findings of this chapter are recapitulated in Section 5.6.

structure

The solutions for these problems can be described by the set of edges used for the topology. They can equally be described by the transmission ranges assigned to the nodes of the network. This alternative rep-

RAP

resentation gave this class of problems its name: Range Assignment Problems (RAPs).

5.2 MATHEMATICAL BACKGROUND

RAPs form a class of optimization problems where an assignment of transmission ranges to wireless nodes is sought such that that the resulting topology satisfies a pre-defined property and the overall energy consumption is minimized [152]. This property could be the connectivity of the wireless network.

one-to-all

There are many interesting RAPs. For example in the Minimum Energy Broadcast (MEB), one node needs to distribute a message to all other nodes in the network. Broadcast routing in wireless ad-hoc networks differs from routing in wired networks. In wireless settings such a broadcast can be achieved by simply adjusting the transmission power of the source to reach all nodes in the ad-hoc network in one hop. However, as mentioned before, intermediate hops may help to reduce the necessary transmission power. All nodes in the network can relay the message, so the broadcast topology is a directed tree spanning the whole network rooted at the first node. Using omnidirectional antennas further brings the advantage of simple local broadcasts, as all nodes within the transmission range can receive the message without additional cost at the sender. This property of the wireless transmission is often referred to as the *wireless multicast advantage*. In the MEB, the leaves of the constructed tree topology do not need to reply to the message, so their transmission range can be set to zero. Asymmetric links are the default in the MEB.

all-to-all

When a reply is needed, e. g. an acknowledgement message, the topology should provide symmetric connectivity. Symmetrical connectivity of the wireless network could be achieved using uni-directional links, by using different routes for each direction of communication. An example was already shown in Figure 2 on page 9. The term *strong connectivity* refers to this type of topology. Although this topology allows every node to reach every other node, it is highly impractical. Many of the low layer protocols in wireless networks require bi-directional links (e. g. CSMA/CA RTS/CTS in IEEE 802.11), so strong connectivity alone is not sufficient in those settings [152]. A topology that uses only bi-directional links for communication represents a *symmetric connectivity*. There are two variants of symmetric connectivity RAPs: The Weakly Symmetric Range Assignment Problem (WSRAP) allows

uni-directional links to be present in the created topology, though they *weakly*
are not used for communication. The Symmetric Range Assignment *symmetric*
Problem (SRAP) requires to dissolve unintentionally created uni-dir- *symmetric*
ectional links or to increase the transmission range of the receiver accordingly to create full bi-directional links [152]. As this increase of the
transmission range may induce other forbidden uni-directional links
which need to be resolved, the SRAP produces higher costs than the
WSRAP. The cost of the SRAP topology can be $\Omega(n)$ times higher than
the corresponding WSRAP topology [152].

The opposite of the MEB is the search for an energy-optimal topology for gathering data from all nodes of the network in one sink node.
Again, the topology is a directed tree, but here the arcs are directed *all-to-one*
towards the root node, which may represent a connection of the wireless network to another network. In this data aggregation tree, the root
node does not need to transmit, but all leaf nodes have to transmit.
There has been exhaustive research on data aggregation in wireless networks. Powerful global view heuristics [85] as well as distributed algorithms [102] have been presented.

In this thesis, we concentrate on broadcast topologies and symmetrically connected topologies. We consider the data aggregation problem to be solved. We will also disregard the SRAP, which forbids uni-
directional links and produces communication costs that can safely be
considered as too high. The following sections give formal definitions
of the considered problems.

5.2.1 *Minimum Energy Broadcast*

The MEB is an $\mathcal{NP}$-hard optimization problem [28, 17]. It is also called
Minimum Power Broadcast (MPB) or Minimum Energy Consumption
Broadcast Subgraph (MECBS). The MEB can be defined as the problem
of finding the broadcast tree $T = (V, E_T)$, a directed spanning tree *definition*
rooted at a source node $s \in V$ in a wireless network $G = (V, V \times V, d)$,
that minimizes the necessary total transmission power $c(T)$ to reach
all nodes of the network:

$$c(T) = \sum_{i \in V} \underbrace{\max_{(i,j) \in E_T} d(i,j)^{\alpha}}_{\text{transmission power of node } i} \tag{5.1}$$

Note that $(i, j) \in E_T$ does not imply $(j, i) \in E_T$, as T is a directed tree.

Using the parent representation as introduced in Section 2.1.2, the cost function for the MEB can also be defined by a parent function $p^T : V \to V$:

parent representation

$$c(T) = \sum_{i \in V} \underbrace{\max_{j:p^T(j)=i} d(i,j)^\alpha}_{\text{transmission power of node } i} \qquad (5.2)$$

In both formulations, the distance function $d : V \times V \to \mathbb{R}^+$ refers to the Euclidean distance and the constant α is the distance-power gradient, which may vary from 1 to more than 6 depending on the environment [86, 152].

In the topologies built by the MEB, each node is required to send only to its farthest child ($\max d(i,j)$), as all other children are then implicitly covered by this transmission. Thus, only the cost for reaching this farthest node has an impact on the objective function of the MEB. Also, the leaves of the tree T do not contribute to the total cost, as they do not need to send to other nodes.

wireless multicast advantage

5.2.2 Minimum Power Symmetric Connectivity

The Minimum Power Symmetric Connectivity Problem (MPSCP) is an $\mathcal{NP}$-hard optimization problem [18, 25]. It is also known as the Strong Minimum Energy Topology (SMET) problem [25] or the Weakly Symmetric Range Assignment Problem (WSRAP) [152]. The MPSCP can be defined as the problem of finding an undirected spanning tree $T = (V, E_T)$ in a wireless network $G = (V, V \times V, d)$, that minimizes the necessary total transmission power $c(T)$ to connect all nodes of the network via bi-directional links:

definition

$$c(T) = \sum_{i \in V} \underbrace{\max_{\{i,j\} \in E_T} d(i,j)^\alpha}_{\text{transmission power of node } i} \qquad (5.3)$$

Again, the distance function $d : V \times V \to \mathbb{R}^+$ refers to the Euclidean distance and the constant α is the distance-power gradient.

In the MPSCP, each node is required to set its transmission range to reach its farthest neighbouring node in the tree. This transmission range determines the necessary transmission power. Note that unlike in the MEB, all nodes including leaf nodes contribute to the total cost, as each node needs to reach at least one other node in the network.

reach all neighbours

The formulation for the MPSCP can be altered to give a formulation for the Minimum Spanning Tree (MST) problem just by exchanging the max with a sum in (5.3). However, while the MST can be solved in polynomial time, the MPSCP is $\mathcal{NP}$-hard [18, 25]. The MST can still be used as an approximation of the MPSCP. It has been shown that the MST already gives a 2-approximation [18, 25].

similarity to MST

5.2.3 Model Limitations

The formulations of the objective functions (5.1), (5.2) and (5.3) are based on a log-distance path loss model [86]. This model describes the relation between distance d and necessary transmission power d^α. In this model, the wireless medium is described solely by one constant: the path-loss exponent α, which is also called distance-power gradient. In an obstacle-free environment it has a value of $\alpha = 2$.

log-distance path loss model

The model assumes that the distance-power gradient remains constant over longer distances. In reality, the environmental conditions for radio transmission may vary largely within the area of a wireless network. Obstacles may affect the distance-power gradient for an individual link. Also, radiation patterns of real antennae are usually not circular. The objective functions can still be used in these cases, if the definition for the distance $d : V \times V \to \mathbb{R}^+$ is changed according to the measured values. As the algorithms presented in this chapter do not depend on special properties of the distance function – it does not need to be metric or even symmetrical – they can be used to optimize the topologies in such networks, given that the necessary transmission powers for all links are provided.

there is no limit

5.3 GLOBAL OPTIMIZATION: MINIMUM ENERGY BROADCAST

In this section we present three global view heuristics for the MEB. The heuristic in Section 5.3.2 combines a Nested Partitioning (NP) algorithm with Local Search and Linear Programming (LP). In Section 5.3.3 an Ant Colony Optimization (ACO) approach is introduced. The steering component of these algorithms is the NP algorithm [155]. NP is a global optimization heuristic that can be used for both stochastic [156] and deterministic problems [134, 158]. The method works by successively partitioning regions expected to contain the best solution into smaller ones, where more concentrated sampling takes place, until a

NP and ACO

singleton, i. e. a region with a single solution, is reached. The algorithm keeps a global view by aggregating the abandoned regions and sampling them. It backtracks to a larger region of the sample space if the abandoned regions are found to be better than the partitioned subregions. This behaviour allows the algorithm to converge to the optimal solution with a positive probability. The LP relaxation is used to find a lower bound for each subregion. If the global best solution is better than this bound, the corresponding subregion will not be sampled. The quality of the samples is further improved by a local search.

ELS

The heuristic in Section 5.3.4 is based on Evolutionary Local Search (ELS), combining Evolutionary Algorithms with Local Search. In this ELS, we use the same Local Search as in the NP and ACO algorithms, which allows us to compare the solution quality as well as the computation time of all proposed approaches.

This section is structures as follows. Section 5.3.1 summarizes related work for global optimization on the MEB. The different global view heuristics are presented in Section 5.3.2, Section 5.3.3 and Section 5.3.4. Results of experiments conducted on a large number of problem instances are given in Section 5.3.5, and Section 5.3.6 summarizes our findings.

5.3.1   Related Work

One of the first approaches for the MEB problem is the Broadcast Incremental Power (BIP) algorithm by Jeffrey E. Wieselthier *et al.* [177, 178].

BIP

This heuristic builds the broadcast tree in a way that resembles Prim's algorithm for building Minimum Spanning Trees (MSTs) [145]. While Prim's algorithm is an exact algorithm for the MST, BIP is an heuristic and does not necessarily find an optimal solution for the MEB. The MST itself can also be used as a heuristic solution for the MEB, but BIP explicitly exploits the wireless multicast advantage and thus produces solutions with lower costs than the corresponding MST solutions. The approximation ratio of MST is known to be 6 for the case of $\alpha \geq 2$ [4], whereas for $\alpha < 2$ the MST does not provide a constant approximation ratio [28]. The approximation ratio of BIP for $\alpha = 2$ is shown to be between $13/3$ and 6 [175].

r-shrink

The BIP heuristic can be further improved by a local search, e. g. *r*-shrink proposed by Arindam K. Das *et al.* in [39]. Here, the transmission power for one node is reduced by *r* steps, cutting off *r* nodes. These nodes will be assigned to other nodes, which increases the latter nodes' transmission power. If the total cost is not reduced, this change is rejected, otherwise it is accepted and the local search is repeated. Experiments have shown that BIP solutions can be improved considerably. The paper [39] only described the case for $r = 1$. Das *et al.* did not implement r-shrink with $r > 1$.

Das *et al.* also developed an ACO algorithm in [37]. They use two   *early ACO*
types of ants: narrow-vision ants and wide-vision ants. Although both types of ants select shorter arcs with a higher probability, wide-vision ants are less greedy in this respect. Das *et al.* state that wide-vision ants are better during the initial exploration phase, while narrow-vision ant are better in the final phase. Results for networks with up to 50 nodes are presented, but no computation times are given. The ACO improved the BIP solutions for their test networks by up to 18 %, but no comparisons to optimal solutions or other heuristics were made. Das *et al.* state that the results could still be improved by including a local search in the ACO, but did not proceed in this direction. Our improved NP algorithm in Section 5.3.3 uses a similar ACO technique and includes local search.

Another improving heuristic called Embedded Wireless Multicast Advantage (EWMA) is presented by Mario Čagalj *et al.* in [17]. Here,   *EWMA*
the transmission power of a node is increased, such that other nodes can be switched off completely. This can be thought of as the opposite of the r-shrink heuristic.

A similar heuristic to the ELS in Section 5.3.4 can be found in [84]. In this paper, Intae Kang and Radha Poovendran present an Iterated Local Search (ILS) heuristic which is based on an edge exchange neighbour-   *ILS*
hood perturbation and the Largest Expanding Sweep Search (LESS) [83], an improved local search by the same authors based on EWMA. This heuristic differs from our ELS as it uses a shrinking operation as mutation and an increasing operation as local search, whereas we propose the opposite. As a result, the broadcast tree in [84] is broken up and repaired in each step of their local search, whereas our heuristic maintains a feasible broadcast tree at all times. Although the idea of increasing the transmission power in the local search is counterintuitive, Kang and Poovendran achieved good results using this heuristic. Unfortunately, when Kang and Poovendran published their ILS, no established standard test instances existed, so comparisons between the heuristics is difficult. Results for 1000 randomly generated instances with up to 20 nodes were presented, where the ILS was compared to optimal solutions. For instances of this size using $\alpha = 2$, the ILS produced an average excess of only 1.1 % over the optimum.

MIP

Two Mixed Integer Programming (MIP) formulations to compute optimal solutions have been presented by Arindam K. Das *et al.* [38] and Roberto Montemanni, Luca Maria Gambardella and Arindam K. Das [124]. While both approaches are based on a network flow model, the MIP from [124] uses an incremental mechanism over the transmission power variables, and is claimed to give better linear relaxations. No experimental results or computation times are given by Das *et al.* in [38], while Montemanni *et al.* successfully solved problem instances with 25 nodes in [124], giving average computation times of about one hour using the commercial MIP solver CPlex [76].

SA

In [124], Montemanni *et al.* also presented a Simulated Annealing (SA) approach for the MEB. However, their results are not competitive to the ILS by Kang *et al.* The SA has one advantage, though: It is very fast. Computation times of only a few seconds are given for problem instances with up to 200 nodes. The SA was able to improve the BIP solutions for problems of this size by only 9.05 %.

survey

A very detailed survey covering these and other heuristics as well as exact algorithms was given by Song Guo and Oliver W. W. Yang in [63]. The survey also covers maximum lifetime problems, as well as multicast variants and scenarios with directional antennae.

ACO with VNS

This survey does not include an improved ACO algorithm, presented by Hugo Hernández *et al.* in [67]. This ACO uses a Variable Neighbourhood Search (VNS) based on r-shrink. Results for problems with up to 50 nodes are presented. As expected, the local search improved the results of the ACO presented by Das *et al.* in [37].

5.3.2   Nested Partitioning Algorithm

general NP

This section presents a Nested Partitioning (NP) algorithm for the MEB. A general NP algorithm works as follows. It iteratively partitions the feasible solution space Θ into disjoint subregions, starting from depth $d = 1$. To evaluate the quality of these subregions, sample solutions are generated. The most promising subregion $\sigma(d)$, i. e. the region containing the best sample solution, is used for the next depth $d + 1$ of the NP. This process of partitioning, sampling and evaluating continues until the most promising subregion consists of only one solution.

To guarantee convergence, the abandoned subregions at $d > 1$ are aggregated in a subregion called the *surrounding region*. The surrounding region is also sampled, so the whole feasible solution space Θ is covered, albeit with different sampling intensities. If the surrounding region is

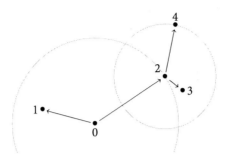

Figure 41: Example of a wireless network with five nodes in the plane. The optimal solution for the MEB is shown, using circles to denote the transmission range, and arcs denoting the message flow. For $\alpha = 2$ the areas of these circles also represent the necessary transmission power.

found to be more promising than the subregions forming $\sigma(d-1)$, the algorithm backtracks to a larger region.

We will use a simple example with five nodes as shown in Figure 41, to explain the NP algorithm for the MEB problem. In this example, *wireless* node 0 is the source node, and the remaining nodes are sorted by their *example* distance from the source. The algorithm executes the following steps: partitioning, finding lower bounds, sampling, determining the promising subregion, and if necessary backtracking. We also introduce a lower bound check which allows the NP to save time by not sampling subregions that are shown to be inferior to the best solution found so far.

Partitioning

The NP for the MEB uses a generic partitioning scheme [155]. The partitioning for the example network is shown in Figure 42. At depth 1, the feasible region Θ is divided into four subregions according to the different nodes the source node transmits to. Allowing the source node 0 to completely switch off its transmission unit would not result in a feasible solution, so this subregion is not generated. Other nodes, however, are allowed to become leaves of the broadcast tree.

In the example, the subregion containing the arc $0 \rightarrow 3$ is found to be the best subregion. At depth 2 the algorithm divides this subregion *splitting*

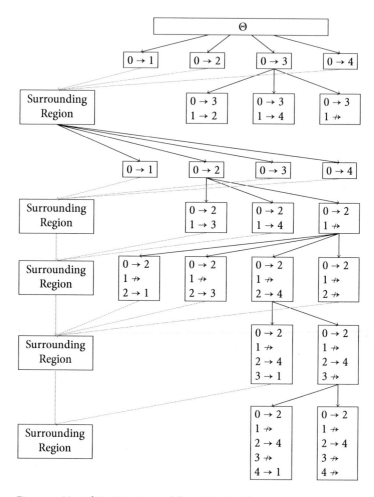

Figure 42: Nested Partitioning workflow. The feasible solution space Θ is divided in smaller regions by iteratively adding arcs to the tree. Only the most promising region is expanded, while all remaining regions form the surrounding region. If the surrounding region proves to be the most promising region, the algorithm backtracks. The algorithm terminates when the promising region cannot be split further.

again into three subregions: One subregion having the arc $1 \rightarrow 2$ as part of the solution, a second having $1 \rightarrow 4$ and a third subregion where node 1 does not transmit to any other node (denoted by $1 \nrightarrow$). In all these subregions, the arc $0 \rightarrow 3$ chosen in depth 1 is also imposed. The surrounding region at this depth is $\Theta \setminus \{0 \rightarrow 3\}$.

In depth 2 of the example, the surrounding region is found to be the promising subregion. Backtracking takes place by reverting to a previous depth, in this case depth 1, by deleting the previously chosen arcs that are not part of the best solution. *backtracking*

To finish the example, assuming that no further backtracking takes place, the remaining steps of the NP algorithm lead to the insertion of the arcs $0 \rightarrow 2$, $1 \nrightarrow$, $2 \rightarrow 4$, $3 \nrightarrow$ and $4 \nrightarrow$. In every step, the abandoned subregions are again aggregated and sampled. This process of partitioning the regions and possible backtracking continues until a singleton, i.e. a region with a single solution, is reached. In this example, the optimal solution as shown in Figure 41 is found. *termination*

Finding Lower Bounds

In order to save calculation time, the NP does not sample regions that are shown to be inferior to the best solution found so far. In the MEB, lower bounds for each subregion can be found using the LP relaxation of the MIP formulation from Montemanni *et al.* [124]. The full MIP formulation is shown in Section A.3.1 in the appendix. In this model, y_{ij} is a binary variable denoting that node i has a sufficiently high transmission power to reach node j. A lower bound can be calculated by relaxing the integrality constraints to $0 \leq y_{ij} \leq 1$. Also, new constraints are added to include all imposed arcs for the respective subregion. In the example, the constraints $y_{01} = 1$, $y_{02} = 1$, $y_{03} = 1$ and $y_{04} = 1$ are used to generate LP formulations for the four subregions at $d = 1$. The lower bounds are calculated using an LP solver, and subregions having higher lower bounds than the best solution found so far S^* can be discarded. *LP relaxation*

A simpler lower bound for each subregion can be found by adding the costs of the imposed arcs. If the cost imposed by the chosen arcs is already higher than the global best solution S^*, then the subregion is not sampled. This reduces the number of subregions to be studied. *simple lower bound*

The best solution S^* found at any stage of the algorithm is passed to the appropriate subregion in the next depth or iteration of the algorithm. Hence, it is never lost and will be returned as the final result of the NP unless a better solution is found.

Sampling

imposed arcs

For subregions resulting from partitioning the most promising subregion, the sampling step begins by forcing the arcs found up to depth $d - 1$ to be part of the solution. For each subregion, the arc that distinguishes it from the other subregions is also chosen to be part of the solution. In the example, the first subregion in depth 1 uses the arc $0 \rightarrow 1$. The first subregion in depth 2 already uses two arcs $0 \rightarrow 3$ and $1 \rightarrow 2$. Sampling the surrounding region is done in a similar way but without imposing any arc on the generated samples.

roulette wheel selection

The rest of the solution is then generated using weighted sampling [155]. The pseudo code for this algorithm is shown in Figure 43. Here, the parameter $q \in [0, 1]$ controls the degree of diversification. With probability q the arc that produces the lowest increase in the total energy consumption is chosen. All other arcs are chosen according to their individual probability using a roulette wheel selection. The probability of choosing an arc is higher for arcs with less additional cost. For $q = 1$ this algorithm matches the BIP construction heuristic [177].

Determining the Promising Subregion

local search

only the best

A local search algorithm is applied to all samples taking into account not to change any of the arcs forced on the subregions. In the NP, we use r-shrink with $r = 1$ as described in [39]. Only the best sample is used to determine the promising subregion $\sigma(d)$. The arcs used by this solution indicate the best subregion.

In case the best sample was generated starting from the surrounding region, the NP checks whether the arcs of this solution follow exactly the chosen arcs from another region. Due to the local search, the sample may have left the surrounding region. The region that contains the best sample, i. e. the lowest feasible energy broadcasting, is selected as the most promising subregion. This region is then partitioned as explained earlier.

Backtracking

If the best solution S^* generated by the sampling is found to be in the surrounding region, backtracking takes place. Backtracking can either be implemented by restarting from the whole feasible region, or by selectively removing chosen arcs and thus returning to a previous depth. The NP algorithms presented in this chapter backtrack to the largest

while $\exists j : \neg covered(j)$ **do**
  for $i = 1 \ldots n$ **do**
    for $j = 1 \ldots n$ **do**
      if $covered(i) \wedge \neg covered(j)$ **then**
        $P_{ij} \leftarrow \left(\text{Extra cost to reach } j \text{ from } i\right)^{-1}$
      else
        $P_{ij} \leftarrow 0$

  $P \leftarrow \dfrac{P}{\sum_{ij} P_{ij}}$ *Normalize probabilities*
  $u \leftarrow \text{RANDOM}[0, 1]$
  if $u < q$ **then**
    $(i^*, j^*) \leftarrow \arg\max_{(i,j)} P_{ij}$ *Select best arc directly*
  else
    $(i^*, j^*) \leftarrow \text{ROULETTEWHEEL}(P)$
  $\text{INCLUDE}(i^*, j^*)$ *Increase transmission range of node i^**

Figure 43: Sampling algorithm for the NP. In each iteration one arc is chosen
to reach new nodes. The probability for choosing an arc is derived
from the additional cost. The increased transmission range may also
cover more than one note due to the wireless multicast advantage.
Once all nodes are covered, the resulting tree is returned.

depth that has the same set of arcs as the best found solution. To achieve   *selective*
this, the backtracking function reduces the depth until all arcs imposed   *backtracking*
in the subregion at that depth are included in the global best solution
S^*. The backtracking function also stops if the whole feasible solution
at $d = 1$ is reached again.

5.3.3   Improved Nested Partitioning

In this section, we present improvements for the original NP algorithm
in Section 5.3.2. This is done by creating a hybrid algorithm combining
Ant Colony Optimization (ACO) with Nested Partitioning and Local
Search.

For these improvements, we use the Max-Min Ant System (MMAS)
presented by Thomas Stützle and Holger H. Hoos in [170], which is a
type of ACO algorithm [47]. These algorithms mimic the collaborat-   *ACO*
ive intelligence behaviour of ants in finding the shortest routes to food

sources. During their search, ants lay a chemical substance, called pheromone, along their routes. Ants follow those pheromone trails which shows them the path from the hive to the food source. The pheromone trails evaporate over time. As they are refreshed by every ant following the trail, shorter routes receive an increasing amount of pheromones, while the pheromone levels on longer routes diminish. In ACOs, this is modelled by having the pheromone persistence depend on the length of a route.

Coding this behaviour of the ants requires the use of a construction graph that represents the considered optimization problem such that ants lay more pheromones along arcs expected to be part of good solutions. Different ACO systems have been proposed that differ mainly in who updates the pheromone trails and how these trails evaporate and are reinforced [47].

ACO in the NP
In the NP algorithm, ACO is used to enhance the quality of the generated sample solutions. ACO has the feature of saving information about the components of good solutions through the pheromone matrix [47]. For the MEB problem, the pheromone matrix is used to save information about good and bad transmission arcs.

reasons for ACO
There are mainly two reasons for using ACO instead of other heuristics within the NP. Firstly, information about good arcs is needed and also provided in the NP algorithm during the sampling step and the selection of the promising region. Secondly, the information provided by the ACO can easily be shared by the different subregions of the NP algorithm.

comparison NP and ACO
Both NP and ACO could be classified as constructive heuristics, as they assemble a solution piece by piece. The ACO repeats this process many times using knowledge from all previous runs which is saved in the pheromones. The NP selects each piece more carefully. It generates random samples for each of the different choices. It picks the best piece based on these sample solutions, adding it to the constructed solution. Thus, the NP constructs the solution by focussing on an iteratively shrinking search space. It can also backtrack, but it loses most of the knowledge from previous runs, except for the best solution. When ACO and NP are combined, the ACO helps to pick good solution components, while the NP reduces the search space, thus speeding up the ACO component.

Max-Min Ant System

The chosen MMAS can easily be implemented within the NP framework since it uses a global pheromone update procedure and the information found in one subregion can easily be applied to ants sampling other subregions by using a global pheromone matrix. The main steps of the MMAS algorithm are: initialization, solution generation, applying a local search and pheromone updating. In the initialization step, all pheromone values $\tau_{i,j}$ are initialized to the maximum allowable value τ_{max}.

Solution Generation

The sample generation algorithm is similar to the code used in the original NP, as shown in Figure 43, with a modification to the probabilistic selection. In the improved NP, the additional cost is not directly translated into a probability value, it only serves as heuristic information η used by the ants:

$$\eta_{ij} = \left(\text{Extra cost to reach } j \text{ from } i \right)^{-1} \qquad (5.4)$$

The pheromone trail between nodes i and j, denoted by $\tau_{i,j}$, is multiplied by this heuristic information. Two parameters a and b are introduced to reflect the importance of pheromone trails and heuristic information. Using an MMAS in this manner, the (not yet normalized) probability value for choosing an arc is calculated as:

$$P_{ij} = \tau_{i,j}^{a} \cdot \eta_{i,j}^{b} \qquad (5.5)$$

probability values

This also means, that the arc with the highest probability value P_{ij} is not necessarily the arc with the lowest additional cost, but an arc selected by the ants based on the additional cost information η_{ij}, the current pheromone trail τ_{ij}, and the exponents a and b.

Pheromone Updating

Every m steps, the pheromone trail is updated using information from the global best solution S^*:

$$\tau_{i,j} = \rho \cdot \tau_{i,j} + \begin{cases} 0.05 & \text{if } (i,j) \in S^* \\ 0 & \text{else} \end{cases} \qquad (5.6)$$

The number of steps m between updates usually stands for the colony size. The parameter $\rho \in [0, 1]$ defines the evaporation rate $1 - \rho$. After each update, it is ensured that all values keep within the boundaries set by $[\tau_{min}, \tau_{max}]$.

NP+MMAS Hybrid

The combination of both algorithms is achieved by modifying the sampling stage of the NP algorithm such that the MMAS is used to generate the samples. In correspondence with the NP algorithm, forcing some arcs that correspond to the different sampled subregions is needed. The pheromone updating function is called every m generated samples regardless of the subregions sampled. This also means that no update is called for several steps of the NP algorithm if the sampling rate is too low or m is too high.

The remaining parts of the NP are left untouched. A simple heuristic is used to find a lower bound for each subregion. If the global best solution is better than this bound, the corresponding subregion will not be sampled. The quality of the generated samples is improved by local search. Again, the r-shrink algorithm is used. It needs to be noted that the convergence of the NP algorithm to the optimal solution depends on the correct selection of the most promising subregion for further partitioning. This in turn depends on the quality of the generated samples and the number of samples taken from each subregion.

5.3.4   Evolutionary Local Search

The MEB heuristic presented in this section is based on Evolutionary Algorithms (EAs) and Local Search (LS) [71]. The general outline of the algorithm is shown in Figure 44. A similar Evolutionary Local Search (ELS) was already used for the Minimum Routing Cost Spanning Tree (MRCST) in Section 3.3.2, or for the Super-Peer Selection Problem (SPSP) in Section 4.3.2.

Representation

As can be seen from the survey [63] and also from the individual heuristics listed in Section 5.3.1, most of the previous local search heuristics for the MEB can be split in two groups based on their neighbourhood structure: Some are based on a tree representation, others are based on

$ar_0 \leftarrow$ INITIALIZATION$()$
$ar_0 \leftarrow$ LOCALSEARCH(ar_0)
$\beta \leftarrow n$ *set mutation rate to problem size*
while $\beta \geq 1$ **do**
   for $i = 1 \ldots \lambda$ **do**
      $ar_i \leftarrow$ MUTATION(β, ar_0)
      $ar_i \leftarrow$ LOCALSEARCH(ar_i)
   $min \leftarrow \arg\min_i \{Z(ar_i)\}$
   if $Z(ar_{min}) < Z(ar_0)$ **then**
      $ar_0 \leftarrow ar_{min}$ *continue search from best offspring*
   else
      $\beta \leftarrow 0.9 \cdot \beta$ *decrease mutation rate*
return ar_0

Figure 44: Evolutionary Local Search for the MEB. In each generation, λ offspring are created using mutation and local search. If an improvement was found, the best solution is used for the next generation.
Otherwise, the mutation rate is reduced by 10 %. The search is terminated once the mutation rate drops below 1, and the best found
solution is returned.

transmission range assignments. A tree representation allows the use
of simple tree operators, whereas the range assignment representation
enables an easier calculation of the total cost. For the ELS, we chose a   *combined*
combination of these two representations, combining their advantages.   *representation*

A solution is represented by a vector ar combining parent assignment and range. For each node i the value $ar(i)$ gives the parent node
of i in the broadcast tree, while $ar(n + i)$ holds the farthest possible
child of i, or i itself if it does not transmit to another node. When a
move is applied during a tree changing operation in the ELS, the range
part is changed as well.

Initialization

The initial solution is created by BIP [177]. Using BIP bears the advant-   *BIP*
age that there is an upper bound for the solutions of the ELS for $\alpha = 2$,
since in this case BIP already gives a 6-approximation for the MEB [4].

The ELS also implements a randomized BIP. While searching for the   *randomized*
next edge to be included in the tree, this version of the BIP adds a ran-   *BIP*
dom value to the costs, such that an edge with slightly larger costs than

the best edge can be chosen. This random version could replace the BIP in the initialization phase to increase the diversity of the population. However, as the ELS uses only one individual, and the randomized BIP is expected to slightly increase the costs of the solutions, the ELS uses an unchanged version of the BIP for initialization. Instead, the randomized BIP is used in the mutation operator.

Local Search

modified
r-shrink

After each step of the Evolutionary Algorithm a local search is applied to further improve the current solution. The ELS uses a modified r-shrink [39]. No levels were computed, instead we apply the local search step to all nodes in the order of their ID.

first best
improvement

During r-shrink, the farthest r children of a node are cut off, reducing the transmission range of their parent node. These nodes have to be assigned to other parent nodes, which increases the transmission ranges of these foster parents. These local search steps are performed for each parent node i. The local search is restarted whenever an improving step was found and applied. This follows a first improvement strategy, however, the best improving move for the first node i that gives an improving move is applied. The local search is thus repeated until no improvement for any node i can be found, i. e. a local optimum has been reached.

The pseudocode for the modified r-shrink is shown in Figure 45. The complexity of r-shrink is hidden in the step of finding the best assignment. Calculating the cost for a given re-assignment is simple: It is the sum of all transmission power increases. Finding the best assignment is harder. The assignment must not connect cut-off nodes to a node in their own subtree, as this would induce a cycle. The assignment also must not connect cut-off nodes to the subtree of other cut-off nodes, if this would create a cycle. Also, the assignment of a node x to a foster parent f_x may reduce the cost for assigning another node y to the same foster parent $f_y = f_x$, thus assigning each individual node to its best foster parent is often not the best choice. In short, the problem of finding the best assignment in the r-shrink local search means solving a smaller version of the MEB. For $r = n$, it is the MEB.

In our implementation, we concentrate on the cases $r = 1$ and $r = 2$, which can be performed in $\mathcal{O}(n^2)$ or $\mathcal{O}(n^3)$.

procedure $\mathrm{LS}(V, d, \alpha, ar, r)$
$n \leftarrow |V|$
for *each transmitting node* i **do**
   $X \leftarrow$ farthest r children of i
                            reduce transmission range to cut off X:
   $ar(n + i) \leftarrow \arg\max_{j \in V \setminus X : ar(j) = i \vee j = i} d(i, j)^{\alpha}$

   $f \leftarrow \arg\min_{f:X \to V} \sum_{x \in X} \mathrm{inc}(f_x, x)$ *find best assignment*

   if *cost is reduced* **then**
      for $x \in X$ **do**
         $ar(x) \leftarrow f_x$ *re-assign nodes*
      for $j \in V$ **do**
         $ar(n + j) \leftarrow \arg\max_{k \in V : ar(k) = j \vee k = j} d(j, k)^{\alpha}$
      restart LS
   else
      $ar(n + i) \leftarrow \arg\max_{j \in V : ar(j) = i \vee j = i} d(i, j)^{\alpha}$ *restore old range*

Figure 45: Modified r-shrink. In each step of the local search, r children of a
transmitting node i are cut off and are re-assigned to other nodes. If
this re-assignment improves the total cost, the local search is restarted. Otherwise, the move is reverted and the next transmitting node
i is tried.

Mutation

Since local search alone will get stuck in local optima, we use mutation
to continue the search. Mutation is done by increasing the transmission power of randomly chosen nodes to reach a randomly chosen fur-   *random*
ther node. The 'gain' of such a move can be calculated as the sum of   *increase*
the power level changes. The mutation changes only the range vector,
not the assignment vector. In these intermediate solutions, the farthest
possible children of some nodes are often not assigned to these nodes.
The local search can then rearrange other nodes to such nodes and thus
make use of the increased ranges.

   In the ELS, several mutation steps are applied in each round. The
number of mutations is adapted to the success rate. The algorithm   *mutation rate*
starts with $\beta = n$ mutations. If no better solution is found in one gener-   *adaptation*
ation, the mutation rate β is reduced by 10 %. This way, the algorithm
can adapt to the best mutation rate for the individual problem and for

the phase of the search. Experience from the MRCST showed that it is favourable to search the whole search space in the beginning, but narrow the search over time, thus gradually shifting from exploration to exploitation.

Population and Selection

one individual only

The ELS uses a population of only one individual. No recombination was implemented. Using mutation and local search, λ offspring solutions are created. The best solution is used as the next generation only if it yielded an improvement. This follows a $(1+\lambda)$ selection paradigm. If there was no improvement in these λ children, the mutation rate β is reduced as described before.

Stopping Criterion

stop when $\beta < 1.0$

The ELS is stopped when the mutation rate drops below $\beta < 1.0$. This value ensures that the neighbourhood of the best solution found by the heuristic is searched especially thoroughly. However, in smaller instances the heuristic often finds the optimum in the first or second generation.

5.3.5 Experiments

A number of experiments have been conducted to check the quality of solutions obtained using the different algorithms. Each experiment was repeated 30 times and average values are used for the following discussion. All CPU times reported in this section refer to a Xeon E5420 2.5 GHz, running Linux. The NP, ACO and ELS algorithms were implemented in C, only the interface to the LP solver CPlex [76] used in the NP needed C++.

Instances

random instances

For our experiments for the RAPs, we used multiple sets of randomly generated test instances. Each set contains 30 instances, where n nodes are randomly located in a $10\,000 \times 10\,000$ grid, using a uniform distribution. Euclidean distance was used and the distance-power gradient was set to $\alpha = 2$ as in an ideal environment. We generated six sets of this type using $n \in \{20, 50, 100, 200, 500, 1000\}$ nodes.

We also created clustered instances with $n \in \{100, 200, 500\}$ nodes bundled in $c = \lfloor \sqrt{n} \rfloor$ clusters, where the cluster centres were placed at a random location in the $10\,000 \times 10\,000$ grid and the node of the clusters were placed at uniformly random locations in a grid of size 1000×1000 around these cluster centres. If a cluster centre was placed close to the border of the original grid, the nodes in this cluster were allowed to leave the original $10\,000 \times 10\,000$ grid.

clustered instances

We named these instance sets after their size: p20, p50, ..., p1000, and for the clustered sets: p100c10, p200c14 and p500c22. The individual instances will be referred to as p20.00, p20.01, ..., p20.29, etc.

naming

As each experiment was also repeated 30 times, this results in 900 experiments for each problem size and each algorithm. Thus, conducting experiments that lasted more than one day was simply impracticable. For the MEB, most of the algorithms took too long when working on the 1000 nodes instances, so these instances are left out in the following discussion.

900 runs each

Obtaining Optimal Solutions

The commercial MIP solver CPlex 10.1 [76] was used together with a modified MIP formulation taken from Montemanni *et al.* [124] to obtain optimal solutions for the problem instances. Please refer to Section A.3.1 in the appendix for details on the MIP formulation. Since the MEB problem is $\mathcal{NP}$-hard, CPlex was only able to provide optimal solutions for the 20 and 50 nodes problems. Depending on the individual problem, CPlex took up to 13 seconds for the 20 nodes problems, and up to 6 days for the 50 nodes problems to find and prove the optimal solution. Tables 35 and 36 in the appendix contain the optimal solutions for these problem instances.

seconds

days

Solving larger problems proved to be too expensive. In one instance of the unclustered 100 nodes set, CPlex took 17 days to arrive at a gap of 26.7 % when it was interrupted by a power outage. The progress reports of CPlex until that point indicated that trying to solve an instance of this size could easily take a couple of months, if not years.

months and years

Evolutionary Local Search

For the ELS, we set the number of offspring solutions to $\lambda = 500$. These settings have proven to be a good choice in preliminary experiments.

settings

In these experiments we also tried larger values of r for the r-shrink local search.

ILS

As a comparison, we used our own implementation of the Iterated Local Search (ILS) presented by Kang and Poovendran [84]. Here, we started from the BIP solution, used the suggested local search (LESS) and mutation operators (edge exchange), but relaxed the termination criterion. Instead of letting the ILS run for only 60 seconds, we allowed it to reach 2000 iterations.

Other comparisons are given by the BIP heuristic and the result of an r-shrink search from this BIP solution. Both can be produced by the ELS, as they are used in the initialization phase of the ELS.

results table

The results for these heuristics are shown in Table 25. The table shows for each algorithm the CPU time, the excess above the optimal or best known solutions, and the number of runs that found these solutions. The best known solutions are taken from all experiments presented in this chapter, including the NP and ACO algorithms shown in Table 26. As the number of problem instances is quite large, each row of these tables shows combined results for all instances of the same size.

n = 20

The ELS using $r = 1$ was able to find the optimum in two thirds of the runs for the 20 nodes problems. Also, the remaining solutions are close to the optimum, with an average excess of 1.39 %.

n = 50

It can be seen that the 50 nodes problems, which are harder to solve for CPlex, are not harder for the ELS. In about one fourth of these runs the ELS still found the optimum, and in the remaining cases the solutions found by the heuristic are still close to the optimum. The average excess over the optimum is increased to 2.45 %.

comparison to BIP

Comparing the results of the ELS against the BIP or BIP+r-shrink shows how much can be gained by the evolutionary approach. The average BIP solutions are more than 20 % higher than the optimal solutions for all considered problem sizes. The average excess of BIP+r-shrink is lower; values of about 10 % are usual. In some of the 20 nodes instances, BIP+r-shrink already finds the optimal solution, but the average of BIP+r-shrink above the optimum is still high: 11.63 %.

comparison to ILS

The results of the ELS are also better than the results of the ILS, when considering the 20 and 50 nodes problems. The ILS found the optimum for the 50 nodes problems in less than one sixth of the runs, and showed an average excess of 1.40 % and 4.01 % for these problems. It is worth mentioning, that the ILS sometimes failed to find the optimum when BIP+r-shrink was successful. Although the mutation operator of the ILS is quite appealing, the local search is counterintuitive and some-

Size	BIP Excess	BIP+r-shrink Excess	ILS Time	ILS Excess	ILS Best	ELS with 1-shrink Time	ELS with 1-shrink Excess	ELS with 1-shrink Best	ELS with 2-shrink Time	ELS with 2-shrink Excess	ELS with 2-shrink Best
20	29.45%	11.63%	0.9 s	1.40%	550	0.2 s	1.39%	602	0.7 s	0.23%	790
50	27.88%	16.22%	18.5 s	4.01%	139	2.2 s	2.45%	235	13.9 s	1.37%	367
100	23.39%	13.89%	203.5 s	5.83%	30	14.8 s	8.54%	13	128.3 s	4.81%	15
200	24.46%	13.64%	2121.0 s	5.23%	12	105.2 s	10.50%	0	1355.4 s	4.56%	44
500	22.32%	12.45%	47384.8 s	2.50%	27	1492.2 s	11.09%	0	42186.2 s	3.44%	1
100c10	23.24%	9.32%	314.7 s	0.95%	146	21.3 s	3.63%	18	161.6 s	1.57%	32
200c14	25.69%	10.93%	4533.8 s	2.77%	0	124.6 s	7.85%	0	2108.5 s	3.42%	24
500c22	23.77%	8.48%	—	—	—	1721.1 s	7.28%	0	63648.3 s	3.00%	18

Table 25: MEB results for LS heuristics. Each line gives average results for all problem instances of the same size. The results of ELS with different r-shrink settings are compared to the BIP construction heuristic, BIP+r-shrink using $r = 1$, and our own implementation of the ILS by Kang et al. [84]. For each heuristic the average CPU time, the average excess above the best known or optimal solution, and the number of runs out of all 900 runs that found this solution is shown.

times fails to make the right decisions. The ILS gains an advantage when larger instances are considered. While the ELS with $r = 1$ produces very high average excess of up to 11 %, the ILS always stays below 6 %. However, the running times of the ILS is much higher than those of the ELS. For the 500 nodes problems, the ILS took more than 14 hours, while the ELS was finished in less than 25 minutes. For the clustered 500 nodes instances, we observed running times of more than two days for the ILS, so these experiments were stopped.

r = 2 To further improve the quality of the ELS solutions, we increased the range of the local search to $r = 2$. This reduced the average excess to about half the values for $r = 1$, but also increased the running times by a factor of 3 to almost 30, depending on the problem size. Also, the number of runs that found the best solutions was increased for all problem sizes. With $r = 2$, the running times are almost comparable to those of the ILS. With the stronger local search, the ELS outperforms the ILS in the uniformly random instances with up to 200 nodes, but still fails in the 500 nodes problems and in the clustered instances.

Nested Partitioning

settings For the original NP, we set the number of samples that are generated for each subregion that passed the lower bounds test to $s = 100$. The probability of directly selecting the best arc bypassing the roulette wheel selection in the sample generation algorithm is chosen as $q = 0.5$. These settings have proven to be a good choice in preliminary experiments. This version of the NP uses CPlex 10.1 [76] and the MIP formulation from Montemanni *et al.* [124] as shown in Section A.3.1 to calculate the lower bounds at each step of the algorithm.

results table The results are shown together with the results of the improved NP variants in Table 26. This table uses the same layout as Table 25; it shows the CPU time, the excess above the optimal or best known solutions, and the number of runs that found these solutions. Again, each row of the table shows combined results for all instances of the same size. As the best known solutions are taken from all experiments presented in this chapter, the values of both tables can be compared directly. The results for the original NP can be found in the column 'NP with CPlex LB'.

n = 20 The NP algorithm was able to find the optimal solutions in almost every run for the 20 nodes problems, Also, the non-optimal solutions

Size	NP with CPlex LB			NP with simple LB			NP with ACO			ACO		
	Time	Excess	Best	Time	Excess	Best	Time	Excess	Best	Time	Excess	Best
20	0.1 s	0.06 %	876	0.2 s	0.00 %	900	0.2 s	0.00 %	900	0.3 s	0.00 %	900
50	3.7 s	3.90 %	159	5.7 s	1.07 %	271	6.1 s	0.19 %	693	8.3 s	0.09 %	785
100	53.3 s	10.00 %	0	78.6 s	4.32 %	42	140.4 s	1.70 %	199	125.7 s	1.26 %	286
200	692.5 s	11.20 %	0	703.4 s	6.48 %	1	2 035.4 s	2.66 %	99	1 873.0 s	2.00 %	172
500	19 333.3 s	12.09 %	0	13 121.7 s	8.63 %	0	57 316.8 s	4.69 %	5	63 456.1 s	1.68 %	20
100c10	83.9 s	3.90 %	30	69.2 s	0.57 %	100	93.1 s	0.08 %	587	131.3 s	0.04 %	709
200c14	1 244.6 s	9.38 %	0	683.2 s	3.92 %	3	2 170.7 s	0.46 %	158	1 778.2 s	0.27 %	174
500c22	28 611.2 s	12.12 %	0	10 126.6 s	7.86 %	0	41 586.8 s	4.08 %	1	60 192.5 s	3.20 %	11

Table 26: MEB results for NP and ACO heuristics. Each line gives average results for all problem instances of the same size. The results of NP using the CPlex lower bound are compared to an improved NP using the simple lower bound, an NP and ACO hybrid, and an ACO heuristic without NP. For each heuristic the average CPU time, the average excess above the best known or optimal solution, and the number of runs out of all 900 runs that found this solution is shown.

are very close to the optimum, with an average excess of 0.06 %. These NP results are also far better than the results of the ILS.

n = 50

For the 50 nodes problems, the NP also found the optimal solutions in about one sixth of the runs. The NP produced a lower average excess than the ILS, but it was outperformed by the ELS with $r = 1$.

n > 50

For the larger instances, the NP produced worse results than the ELS with $r = 1$, which was already outperformed by the ILS. With an average excess of more than 12 % and running times of about half the values of ELS with $r = 2$, this version of the NP is not competitive.

CPU time

The observed CPU times for the different problem sizes match the expected average time complexity of $\mathcal{O}(n^4)$. In each depth $1 \ldots n$ of the NP, n subregions are sampled, where the weighted sampling step uses $\mathcal{O}(n^2)$ time. However, when backtracking takes place more often, the time complexity will increase.

bad example solution

To give an illustration of the generated solutions, Figure 46 displays the optimal solution and the worst solutions found by the ELS and the NP for problem p50.19. The ELS showed its worst performance on p50.19 in terms of average excess above optimum, although it found the optimum in every third run. In the remaining runs, it often got stuck in an especially bad local optimum, as shown in the figure. While in the optimal solution node 3 sends to the majority of the nodes, this role is given to node 40 in the worst solution of the ELS. The ELS simply cannot escape this local optimum, as it would have to assign most of these nodes to other nodes during the mutation. If the range of node 40 cannot be decreased after this mutation, the r-shrink local search would simply assign all these nodes back to node 40, as this does not increase the total cost. As NP allows backtracking, it can revert such star topologies easier than the ELS, once a better solution has been found in the surrounding region. The main difference between the optimal solution for this network and the worst solution found by the NP can be found in the bottom part of the network, where the path from node 39 was not found and a path from node 29 is used instead.

Improved Nested Partitioning

different lower bound

For the improved NP, we changed the lower bound calculation. Instead of having CPlex solve an LP relaxation, we use the simple lower bound by adding up the costs for all imposed arcs. This already improved the algorithm. The original NP spend a very large portion of its running time in the LP solver. By switching to the simpler lower bound, the

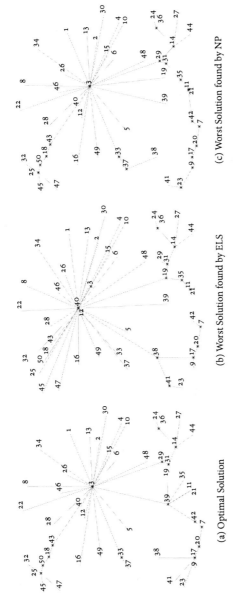

(a) Optimal Solution

(b) Worst Solution found by ELS

(c) Worst Solution found by NP

Figure 46: Optimal and sub-optimal solutions for the MEB in p50.19. The source is node 3, and transmitting nodes are highlighted.

NP produces slightly worse results, but takes much less time. Increasing the sample rate is already enough to improve the solution quality beyond the original NP.

pure NP

The column 'NP with simple LB' in Table 26 gives the results for a pure NP using the simple lower bound, an increased sample rate of $s = 500$, and $q = 0.9$. Better results could be found with higher sampling rates, but the computation times would increase. The NP results are already quite good: All runs on the 20 nodes problems found optimal solutions, and an average excess of 1.07 % for the 50 nodes problems is reached in only 5.7 seconds. Also, the average excess for the 100 nodes problems, both clustered and uniformly distributed, is lower than that of the ELS with $r = 2$.

For the larger problems, the improved NP still outperforms the ELS with $r = 1$ with an exception at the clustered 500 nodes problems, but is outperformed by the ELS with $r = 2$ and the ILS.

Nested Partitioning + Max-Min Ant System

MMAS settings

When using MMAS, the pheromone update function was called after every $m = 50$ samples. The evaporation rate of the pheromones was set to 2 % ($\rho = 0.98$), and an additional 0.05 pheromone trail was added on the edges that are used in the best solution known at that point in the algorithm according to (5.6).

parameter study

An extensive parameter study was carried out to determine best values for the sample rate s, the influence gradients a and b (influence of pheromones and heuristic information), and the probability q to directly select the best arc when generating the sample solution. Tables 38 and 37 in the appendix show the results of this parameter analysis. Based on these results, we used the following settings for our experiments: $a = 1$, $b = 2$, $q = 0.2$.

NP+ACO

The influence of the pheromone trail in the NP algorithm is quite remarkable. The column 'NP with ACO' in Table 26 gives the results. The calculation time is increased by the additional effort for generating sample solutions. They are still comparable to those of the ILS. The NP with ACO finds more optimal or best known solutions than the pure NP. It reduces the average excess of the 20, 50, and the clustered 100 and 200 nodes problems to values below 0.5 %. It outperformed all previous heuristics, except for the 500 nodes problems, where ILS or ELS with $r = 2$ produced slightly better average results.

Max-Min Ant System

The impact of the pheromone information on the quality of the results suggested to conduct experiments with a pure MMAS. In these experiments, the NP partitioning scheme was deactivated. Instead, only the sample solution generation and the pheromone update procedure from the NP with ACO was used. Again, the pheromone update function *settings* was called after every $m = 50$ samples, and the evaporation rate of the pheromones was set to 2 %. The settings for the influence gradients a and b as well as for the parameter q remained the same: $a = 1$, $b = 2$, $q = 0.2$. We also created the same number of samples as in a run of the NP when no backtracking occurred: $s = 500 \cdot n$.

Results for this pure MMAS are shown in column 'ACO' in Table 26. This heuristic found the optimal or best known solutions more often *ACO* than any of the other heuristics. It produced average gaps below 2 %, outperforming the ILS, the ELS with both $r = 1$ and $r = 2$, and even the improved NP versions. There is one exception: For the clustered 500 nodes problem, the ELS with $r = 2$ produced a slightly smaller average excess, and found more best known solutions than this MMAS.

Best Known Solutions

The best known solutions have been found most frequently by the ACO or the ACO-enhanced NP. However, for the largest clustered problems, *p500c22* p500c22, the ELS using $r = 2$ found best known solutions for most of the instances. For this problem set, the best known solution has only been found in one of the 30 runs for each instance.

For the unclustered problems of this size, p500, the best known solutions have been found in multiple runs. The ILS from Kang *et al.* found *p500* the best known solution for 5 instances. One of these best known solutions was found in 23 out of 30 runs. This could be taken as an indication that this solution is optimal. The NP using ACO also found the best known solution for 5 instances. The ELS using $r = 2$ found the best known solution for one instance. For the remaining 19 instances, the ACO found the best known solutions.

5.3.6   Discussion

This section compared different heuristics for the global view optimization of the Minimum Energy Broadcast (MEB). In addition to the familiar approach of using a form of iterated or evolutionary local search, we also looked at a relatively new type of heuristic: Nested Partitioning (NP). The original NP algorithm and results for the 20 and 50 nodes problems were already published together with Sameh Al-Shihabi in [160]. As the results of this algorithm using the commercial LP solver CPlex were rather weak, we proposed a number of improvements. Using a simpler lower bound instead of the time consuming LP relaxation allowed the NP to spend more time finding good sample solutions. A further improvement was the introduction of an Ant Colony Optimization (ACO) approach, also joint work with Sameh Al-Shihabi. However, as the NP algorithm restricts the sample generation and subsequent local search inside the defined sub-regions, an approach solely based on ACO brings even better results.

NP

There are a number of parameters controlling the NP algorithm. As with nearly all heuristics, NP can be fine-tuned to produce better results. One direction of future research is an adaptive sampling rate. In the current version, we used the same sampling rate for all sub-regions of the NP. A possible improvement is to increase the sampling rate for the more promising sub-regions. The sampling rate could also be decreased with increasing depth, as the sub-regions are getting smaller while the NP fixes more and more arcs. Another direction is to use information from the LP relaxation to guide the search to more promising solutions. This could be done almost as in an ACO, using the LP relaxation values as probability values.

future NP

The ELS approach using the established r-shrink local search [39] also found good solutions in short time. A similar ELS using a modified r-shrink local search and results for the 20 and 50 nodes problems were already published in [182]. The ELS presented in this chapter outperforms the original NP version. The ELS can still be improved by using a stronger local search. The results using 2-shrink almost match the results of the improved NP without ACO. However, as the mutation operation in the ELS is blind to the problem instance and the current solution, it often destroys already optimal parts of this solution and either results in longer running times for the subsequent local search or in weak solutions. The mutation rate adaptation helps to reduce this effect as it decreases the number of changes when the solution quality increases.

ELS

Including information about good arcs in the mutation operator already leads to the ACO approach. In a way, this can be seen as a simple ELS using the r-shrink local search and a more sophisticated muta-

ACO

tion based on pheromone information. Using ACO, we were able to decrease the gap to the optimal solutions to satisfying values.

When we were still conducting experiments with the MMAS, Hugo Hernández, Christian Blum and Guillem Francès published an ACO algorithm in [67] using a Variable Neighbourhood Search (VNS) based on r-shrink. They improved upon the results presented in this thesis, using our 20 and 50 nodes test instances. The VNS approach allows them to improve the results of 1-shrink without increasing the computation times too much. In the VNS, the higher level r-shrink local searches can already start from very good solutions, which were optimized by 1-shrink and all subsequent lower searches. As r-shrink with $r = n$ also poses an $\mathcal{NP}$-hard problem (the same problem as the MEB), they used a parameter to limit the CPU time for the higher level r-shrink searches. An average excess of 0.004 % over the optimal solutions was reported for the 50 nodes set. Although this is an independent work – they were inspired by our original NP approach – their ACO can be seen as the next logical step from the ACO presented in this thesis.

Before our work on the NP for the MEB, comparisons to other heuristics were difficult, as each publication used another set of randomly generated test instances. Since our test instances were proposed in [160] and have been publicly available, they have been used by other researchers.

During our experiments, no best known solution for problems with 500 nodes was found frequently by more than one heuristic. However, for the smaller instances, this happened very often. Although we cannot prove the optimality of the best known solutions for problems with 100 nodes or more, we still believe that the best known solutions found for all 100 nodes instances are optimal, and those for the 200 nodes instances are very close to the optimal solutions. There is probably still some room for improvements for all 500 nodes problems.

Christian Blum

problem instances

optimality

5.4 GLOBAL OPTIMIZATION: SYMMETRIC CONNECTIVITY

In this section we present an Iterated Local Search (ILS) heuristic for the Minimum Power Symmetric Connectivity Problem (MPSCP) that works on a global view of the wireless network. The local searches are based on changes in a tree that builds the symmetric connectivity topology. Some of these local searches resemble local searches that have been successfully used in Section 3 for the MRCST, however, the calculation of the gains of these moves needs to be adapted to the wireless setting. In the RAPs the length of an included edge, i. e. the distance between two nodes, does not directly contributes to the cost. Thanks to the wireless multicast advantage, a node can send one transmission and reach multiple nodes without additional cost.

This section is structured as follows. In Section 5.4.1, we first give an overview of related work for global view optimization of the MPSCP. The ILS and the used local searches are presented in Section 5.4.2. Experiments with this ILS and the results are given in Section 5.4.3. Section 5.4.4 summarizes our findings.

5.4.1 Related Work

introduction

The MPSCP was first formulated by Gruia Călinescu *et al.* in [18] as a variant of another RAP, which searched for strong connectivity only [27, 88]. Many of the findings were transferred to the MPSCP, e. g. the

$\mathcal{NP}$-hard

proof for $\mathcal{NP}$-hardness for general graphs and for Euclidean graphs in $\mathbb{R}^2$. Călinescu *et al.* also show that the MST is a 2-approximation for the MPSCP, based on a similar result for the strong connectivity

2-approximation

variant in [88]. This approximation ratio is shown to be tight using the example from Figure 47. They also give an approximation scheme based on k-restricted decomposition yielding an approximation ratio of $1 + \ln 2 + \varepsilon \approx 1.69 + \varepsilon$.

rediscovery

Xiuzhen Cheng *et al.* independently formulate the same problem in [25]. They call it Strong Minimum Energy Topology (SMET), and emphasize that the SMET or MPSCP is different than the strong connectivity RAP and provide a new proof of $\mathcal{NP}$-hardness by reduction from the vertex cover problem (GT1 in [57]). They also rediscover the MST as a 2-approximation, and present a new greedy heuristic, later called

IPP heuristic

Incremental Power: Prim (IPP). This heuristic builds the tree in a way that resembles Prim's algorithm for building MSTs [145]. Since IPP explicitly exploits the fact that nodes can reach closer nodes without additional cost when they already have to send to a farther node, it produces solutions with lower costs than the corresponding MST solutions.

Two local search heuristics to improve these construction heuristics were presented by Ernst Althaus *et al.* [2, 3]. In the Edge Switching

ES heuristic

Search (ES) heuristic, edges from the tree are replaced by non-tree edges re-establishing the connectivity. In the Edge and Fork Switching Search (EFS) heuristic, not only edges, but also pairs of edges shar-

EFS heuristic

ing one node, so-called forks, are inserted in the tree, and the result-

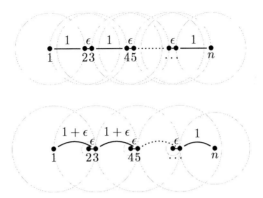

Figure 47: Tight example for performance ratio of MST approximation. The MST shown at the top needs a total power of $n \cdot 1^{\alpha}$, each node needs to send over the larger link. The optimal assignment shown at the bottom needs $n/2 \cdot (1 + \varepsilon)^{\alpha} + (n/2 - 1) \cdot \varepsilon^{\alpha} + 1^{\alpha} \overset{\varepsilon \to 0}{=} n/2 + 1$. Only about half the nodes need to send over a larger link. This example is taken from [18].

ing cycles are cut again by removing other edges. In both heuristics, the largest reduction in cost is chosen, and the process is repeated until a local optimum is reached. Forks were already used in [18] as 3-restricted decompositions, but the EFS produces better results. Average improvements over the MST of up to 6 % are achieved. Another construction heuristic, later called Incremental Power: Kruskal (IPK), that builds the tree in a way that resembles Kruskal's algorithm for building MSTs [94] is also used as a comparison. The authors also try filtering out edges to reduce computation complexity, and concentrate on Delaunay edges only. As this may also filter out edges that are part of the optimal solution, the results using the filtered local searches are weaker.

IPK heuristic

Delaunay filter

Joongseok Park *et al.* give an experimental survey in [138], describing different construction heuristics and local search heuristics. They name the greedy heuristics based on Prim's algorithm [145] or Kruskal's algorithm [94] Incremental Power: Prim (IPP) and Incremental Power: Kruskal (IPK), respectively, and also present a new local search Double Edge Switching Search (ES2) that applies the best double edge switch. They further show that the approximation ratio of 2 is not reduced by these heuristics. In the experiments, problem instances with up to 100

IPP and IPK names

ES2 heuristic

nodes are used, but no computation times are provided. Average improvements over the MST of up to 6 % are presented. For the ES2 local search, the average improvements are claimed to be as high as 14 %. We were unable to reproduce these values using similar random instances. In our instances, the *optimal* solution is about 6-7 % below the MST. However, their instances may contain some properties that are unknown to us.

MIP formulation

Several Mixed Integer Programming (MIP) formulations have been presented to compute optimal solutions. Ernst Althaus *et al.* give an MIP formulation in [2, 3] that uses an exponential number of inequalities to ensure connectivity and to forbid cycles. A cutting plane algorithm is used to find optimal values for problem instances of up to 40 nodes (or up to 60 nodes when only Delaunay edges are considered).

two MIP formulations

Roberto Montemanni and Luca Maria Gambardella [123] give two MIP formulations based on incremental power. In the first formulation they avoid the problem of an exponential number of inequalities by defining n^3 flow variables. The second formulation is incorporated in an exact solver EX2, which iteratively solves the MIP and adds necessary constraints. The authors also present a filtering technique (see

filter technique

Section 5.4.2), that considerably reduces the computation time but does not affect correctness.

computation times

Both Althaus [2, 3] and Montemanni [123] give results of experiments conducted on problem instances with up to 40 nodes. While Althaus *et al.* take more than an hour for each of these instances, Montemanni *et al.* can solve an instance of this size in less than two minutes. However, calculation times increase exponentially with the network size for both algorithms.

approximation algorithms

In many of these papers, approximation algorithms have been presented, reducing the approximation ratio from 2 for the MST, IPP and IPK down to $1 + \ln 2 + \varepsilon$ [18] and $5/3 + \varepsilon$ [2, 3].

difference to MEB

Heuristics for the MEB cannot be directly reused for the MPSCP. The MEB allows leaves of the broadcast tree to completely switch off their radio transmitter, yet they are still connected to the network. In the MPSCP, each participating node in the wireless network needs to transmit to at least one other node. Links are established only if both communication partners increase their transmission ranges to sufficient levels to reach the other node. Construction heuristics as well as local search and mutation operators in an EA need to take the additional costs for these symmetric links into account.

$T \leftarrow \text{Initialization}(E)$
$F \leftarrow \text{Filter}(E, T)$
$T \leftarrow \text{LocalSearch}(E, T, F)$
repeat
    | $F \leftarrow \text{Filter}(E, T)$
    | $T_{\text{new}} \leftarrow \text{Mutation}(E, T, F)$
    | $T_{\text{new}} \leftarrow \text{LocalSearch}(E, T_{\text{new}}, F)$
    | **if** $c(T_{\text{new}}) < c(T)$ **then**
    | $\lfloor$ $T \leftarrow T_{\text{new}}$
until *Termination* ;

Figure 48: The Iterated Local Search (ILS) for the MPSCP. After initialization
and mutation a local search is applied. Local search and mutation
are enhanced by a filter technique that marks edges that cannot be
part of an optimal solution.

5.4.2 *Iterated Local Search*

The MPSCP heuristic presented here is based on ILS [103]. It operates
on a global view of the wireless ad-hoc network. It also incorporates a
filtering technique first presented in [123] to exclude edges that cannot
be part of the optimal solution. The general outline of the algorithm
is shown in Figure 48. The following paragraphs describe the different
parts of the ILS.

Representation

This ILS uses the same representation as the ELS for the MEB. Each
node stores its transmission range as defined by the farthest neighbour
to be reached. The tree topology is represented by storing the parents
of each node. Unlike the MEB, there is no defined source node in the   *parent notation*
MPSCP. The ILS arbitrarily chooses a random node to be the root of
the tree. This root is allowed to change. In fact, it is changed a number
of times during a single local search step. As the parent pointers already
define the path to the root, the ILS implementation also uses them to
show the paths to arbitrary other nodes. This is done by simply moving
the root to this node, as already illustrated in Figure 1 on page 7.
    The direction of the arcs has no influence on the mutation or local   *direction is*
search. It is also not used for the calculation of the objective, as this is   *irrelevant*

defined by the sum of all transmission ranges, which are stored separately.

Filter

In order to reduce the number of edges to be considered by the heuristic, the ILS applies a filtering technique as described in [123]. Since each node needs to have a transmission range at least high enough to reach its nearest neighbour, a very weak lower bound can be calculated. The filter then checks whether adding an edge $\{i, j\}$ increases this cost beyond an upper bound determined by the best solution found so far *filter out edges* by the heuristic. Such edges are marked and are not used in the following operations. In the ILS the filter is applied for a first time after the initialization.

During the run of the ILS, whenever a better solution is found, more edges can be filtered out. This is done by storing the induced costs for each edge and comparing these costs to the current best known solution S^* when checking whether an edge is included in the filter F.

Surprisingly, in some cases a larger edge might be allowed, although smaller edges starting from the same node are filtered out. This is the case, for example, when this node is the nearest neighbour of a remote node.

Initialization

The initial solution for the ILS is created using MST or IPP. Using MST and IPP bears the advantage that there is an upper bound for the solutions of the Iterated Local Search heuristic, since both heuristics already give a 2-approximation for the MPSCP [18, 138]. Although IPP tends to produce better solutions, in some cases its solutions are slightly *best of MST* worse than the MST. The ILS simply picks the better solution as initial*and IPP* ization.

Local Search

After initialization and mutation, a local search is applied to further improve the current solution. For the ILS, we use a modified Edge Switching Search (ES), as well as an Edge and Fork Switching Search (EFS) as defined in [2, 3]. We also use a faster Subtree Moving Search (ST), which was already used for the MRCST in Section 3.3.2. Each local search heuristic follows a best improvement strategy. The local search is restarted whenever an improving step was found, until a local optimum has been reached.

The local searches used in the ILS are based on edge exchanges. They can be implemented as insert search or deletion search, where an insert search version would first insert the new edge and then find the best *insert or delete?* edge to be removed, while a deletion search version would first remove an edge an then find the best edge to reconnect the partitioned tree. As both versions lead to the same local optima, we picked the fastest implementations for the ILS. For most of the considered local searches, the fastest implementation was based on insert search. A similar observation was already made for the MRCST in Section 3.3.2, where the insert search version reduced the expected time complexity for the local search.

The pseudocode for the filtered Edge Switching Search (ES) is shown in Figure 49. It works by inserting an edge $\{i, j\}$, which creates a cycle. *ES* From this cycle the edge whose removal gives the best savings is removed. In our implementation, the simple graph operations such as inserting or removing an edge take constant time, while following the path defined by the tree edges takes linear time based on the path length. The time complexity of one step of ES is $\mathcal{O}(n^3)$, as there are $\mathcal{O}(n^2)$ edges $\{i, j\}$ to be considered, and the path between any two nodes $i, j \in V$ contains at most $n - 1$ edges. As the path length can be expected to be in $\mathcal{O}(\log n)$ in balanced trees, the expected time complexity of one step of ES is reduced to $\mathcal{O}(n^2 \cdot \log n)$.

Calculating the possible savings for every edge is done at the beginning of each step of the ES. Only edges adjacent to nodes i and j need recalculation, since the newly included edge $\{i, j\}$ changes the precalculated savings. The local search was also modified not to insert edges that cannot be part of an optimal solution and are therefore filtered out (F) in a pre-processing step. The unmodified ES can be simulated by setting $F = \varnothing$.

Another local search used in the ILS is the Edge and Fork Switching Search (EFS). This search tries to insert either a single edge or a fork. A *EFS* fork consists of two adjacent edges $\{a, b\}$ and $\{a, c\}$. As inserting two edges creates two cycles, one edge from each cycle needs to be removed to repair the tree. However, the cycles may overlap, as Figure 50 shows. There are three paths that need to be checked: $P_{a \to y}$, $P_{b \to y}$ and $P_{c \to y}$. Removing two edges from the same path would cut the tree. Instead, two edges from two different paths need to be removed.

procedure ES(E, T, F)

repeat

> $g^* \leftarrow 0$
>
> $(g_{\{k,l\}}) \leftarrow$ PRECALCULATESAVINGS(E_T)   *savings for removing edges*
>
> **for** *each edge* $\{i,j\} \in E \setminus (E_T \cup F)$ **do**
>
> > $P_{i \rightarrow j} \leftarrow$ Path from i to j in E_T
> >
> > $c_{i,j} \leftarrow$ INSERT($\{i,j\}, E_T$)   *increases ranges of i and j*
> >
> > $\{x,y\} \leftarrow \arg\max\limits_{\{k,l\} \in P_{i \rightarrow j}} g_{\{k,l\}}$
> >
> > **if** $g_{\{x,y\}} - c_{i,j} > g^*$ **then**
> >
> > > $g^* \leftarrow g_{\{x,y\}} - c_{i,j}$
> > >
> > > $(i^*, j^*, x^*, y^*) \leftarrow (i, j, x, y)$
> >
> > REMOVE($\{i,j\}, E_T$)   *restores ranges of i and j*
>
> **if** $g^* > 0$ **then**
>
> > INSERT($\{i^*, j^*\}, E_T$)
> >
> > REMOVE($\{x^*, y^*\}, E_T$)

until $g^* = 0$;

Figure 49: Edge Switching Search (ES), implemented as insert search. Filtered edges F are not inserted in the tree. Including edge $\{i, j\}$ in the tree increases the cost of this solution by $c_{i,j}$. From the created cycle the edge that gives the best gain g is determined. If the best exchange yields a positive total gain, the move is applied. The process continues until no more improvements can be found.

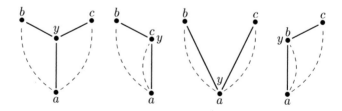

Figure 50: Cycles induced by inserting a fork consisting of the edges $\{a, b\}$ and $\{a, c\}$. In these simple examples, the induced cycles are (a, b, y, a) and (a, c, y, a). The cycles usually overlap on the path from a to a node y, where both cycles meet, as the left example shows. The remaining three examples show the cases where node y is the same node as a or b or c.

The EFS finds the best possible exchange by comparing the additional cost for inserting the fork or edge to the energy saved by removing the most expensive two edges from the cycle. This can be done in $\mathcal{O}(n^4)$, as the pseudo-code in Figure 51 shows. The expected time complexity is in $\mathcal{O}(n^3 \cdot \log n)$, as the expected path lengths is in $\mathcal{O}(\log n)$.

Note that calculating the possible savings for removing an edge can be done in an outer loop, but these savings will change for all links adjacent to the nodes a, b, c and y, as their transmission range is changed during the search. These adjustments are not explicitly stated in the pseudo-code.

A third local search is the Subtree Moving Search (ST). This search was implemented as deletion search. In the ST, an edge $\{i, j\}$ is removed from the tree. The node i becomes the root node of its subtree. This subtree is reconnected with the main tree by connecting i to another node in the main tree. Again, the best exchange is applied if it yields a positive gain. This best exchange can be found in $\mathcal{O}(n^2)$ time, as the pseudo code in Figure 52 shows.   *ST*

A simple Variable Neighbourhood Search (VNS) was implemented using two of these local searches. It works by sequentially applying ES and EFS to the intermediate solutions. This gives a speed-up for the EFS, as it can start from already good solutions. If EFS can improve the local optimum of the ES, this new solution is returned. The quality of these solution is expected to be weaker compared to applying EFS on the initial tree, as the ES could lead the search back to the attraction basin of another EFS local optimum.   *simple VNS*

More expensive local searches are also possible. For example, the EFS could be extended to insert two non-adjacent edges in the tree and remove one edge from each of the created cycles, thus yielding the ES2 local search from [138] with time complexity $\mathcal{O}(n^6)$.

Mutation

Since local search alone will get stuck in local optima, the ILS uses mutation to continue the search. The mutation operator should change the solution enough to leave the attraction basin of the local optimum, but it should also avoid changing the solution too much and destroying already promising structures. Another criterion for the mutation operator is its relation to the local search. The mutation operator should not be easily reversible by the local search. In the ILS the following two

procedure EFS(E, T, F)
repeat

> $g^* \leftarrow 0$
> PRECALCULATESAVINGS(E_T)   *Savings for removing edges*
> **for** *each pair* $(a, b) \in E : a \neq b \wedge \{a, b\} \in E \setminus (E_T \cup F)$ **do**
>
> > $P_{b \to a} \leftarrow$ Path from b to a in E_T
> > $c_{a,b} \leftarrow$ INSERT($\{a, b\}, E_T$)   *increases ranges of a and b*
> > **for** *each* $c : c \geq b \wedge \{a, c\} \in E \setminus (E_T \cup F) \vee c = b$ **do**
> >
> > > $P_{c \to a} \leftarrow$ Path from c to a in $E_T \setminus \{a, b\}$
> > > $c_{a,c} \leftarrow$ INSERT($\{a, c\}, E_T$)   *increases ranges of a and c*
> > > $P_{y \to a} \leftarrow P_{b \to a} \cap P_{c \to a}$ *largest common path from a to b or c*
> > > $y \in \{a, b, c\}$ *is possible*
> > > Find best edges on each path $P_{y \to a}$, $P_{b \to y}$ and $P_{c \to y}$
> > > Check whether removing two edges around y saves more
> > > $(\{i, j\}, \{k, l\}) \leftarrow$ pair of edges that saves the most
> > > **if** *saved costs* $- c_{a,b} - c_{a,c} > g^*$ **then**
> > >
> > > > $g^* \leftarrow$ saved costs $- c_{a,b} - c_{a,c}$
> > > > $(a^*, b^*, c^*, i^*, j^*, k^*, l^*) \leftarrow (a, b, c, i, j, k, l)$
> > >
> > > REMOVE($\{a, c\}, E_T$)   *restores ranges of a and c*
> >
> > REMOVE($\{a, b\}, E_T$)   *restores ranges of a and b*
>
> **if** $g^* > 0$ **then**
>
> > REMOVE($\{i^*, j^*\}, E_T$)
> > REMOVE($\{k^*, l^*\}, E_T$)
> > INSERT($\{a^*, b^*\}, E_T$)
> > INSERT($\{a^*, c^*\}, E_T$)

until $g^* = 0$;

Figure 51: Edge and Fork Switching Search (EFS), implemented as insert search. Filtered edges F are not inserted in the tree. Including an edge in the tree creates a cycle. If two edges are inserted, the created cycles could overlap. Node y that connects these cycles is determined. From all three paths $P_{y \to a}$, $P_{y \to b}$, $P_{y \to c}$, only two edges are chosen to be removed. The best exchange is applied and the process continues until no more improvements can be found.

procedure $ST(E, T, F)$ *F is ignored*
repeat
   $g^* \leftarrow 0$
   for *each edge* $\{i, j\} \in E_T$ **do**
      $g \leftarrow \text{REMOVE}(\{i, j\}, E_T)$ *decreases ranges of i and j*
      for *each* $v \in V \setminus \text{SUBTREE}(i)$ **do**
         $c \leftarrow \text{INSERT}(\{i, v\}, E_T)$ *increases ranges of i and v*
         if $g - c > g^*$ **then**
            $g^* \leftarrow g - c$
            $(i^*, j^*, v^*) \leftarrow (i, j, v)$
         $\text{REMOVE}(\{i, v\}, E_T)$ *restores ranges of i and v*
      $\text{INSERT}(\{i, j\}, E_T)$ *restores ranges of i and j*
   if $g^* > 0$ **then**
      $\text{REMOVE}(\{i^*, j^*\}, E_T)$
      $\text{INSERT}(\{i^*, v^*\}, E_T)$
until $g^* = 0$;

Figure 52: Subtree Moving Search (ST), implemented as deletion search. The filter F is not used in this version. Removing an edge $\{i, j\}$ from the tree splits the tree. The subtree rooted at node i is reconnected to the main tree using the least expensive edge. If the best exchange yields a positive gain, the move is applied. The process is repeated until no more improvements can be found.

mutation operators are used: random range increase and random edge exchange.

In the first mutation operation, the transmission power of a randomly chosen node is increased to reach a randomly chosen further node. All newly reached nodes adjust their transmission power to establish bi-directional links to the first node. Unnecessary connections, i. e. those that would introduce a cycle, are cut without adjusting the transmission power. The local search can later adjust these higher transmission power settings or use them to reach other nodes. The cost of the solution increases according to the sum of the power level changes. Since this mutation operation could result in star-like topologies, we applied the same filters as for the local search. Also, with 20 % probability we increased the transmission range of the selected node to the largest level allowed by the filters.

random
increase

edge exchange
In the second mutation operation, a random edge is removed from the tree and another random edge is inserted to reconnect the tree. The transmission ranges of the involved nodes are adjusted accordingly. However, new connections that could be established because of the increased ranges are not inserted in the tree. This can be done later by the local search heuristics. As a single step of this mutation operator is easily reversed by almost all of the used local searches, we applied this mutation multiple times.

Population and Selection

one individual only
The ILS uses a population of only one individual. Again, there is no recombination. Using only mutation and local search, one offspring solution is created. The offspring is used as the next generation if it yields an improvement. This follows a $(1 + 1)$-ES selection paradigm.

The ILS could be expanded to a full ELS or a Memetic Algorithm (MA) by using larger populations and recombination. However, the experiments in Section 5.4.3 show that the ILS approach already gives satisfying results.

Termination Criterion

fixed number of iterations
We stopped the ILS after 200 iterations. For the smaller instances ($n \leq 50$), this setting is already too high. However, for the larger instances ($n \geq 500$) and especially for the weakest local search (ST), this setting should be increased in order to find optimal solutions. Because of the large running times of the stronger local searches, we increased the number of iterations to 2000 only for the ST local search.

As we wanted to compare the local searches using the same settings for each experiment, we refrained from using heuristic termination detection.

5.4.3   Experiments

The experiments for the MPSCP were conducted using the same settings as for the MEB. Each experiment was repeated 30 times, the same set of problem instances was used. Calculation times again refer to the a 2.5 GHz Intel Xeon, and the algorithms were implemented in C.

Obtaining Optimal Solutions

The commercial MIP solver CPlex 10.1 [76] was used together with our own implementation of the exact algorithm EX2 from [123] to obtain optimal solutions for the problem instances. EX2 uses the same filtering technique as described in Section 5.4.2. We used the cost of the best known solution found in any of our experiments as an upper bound and filtered out all edges that would induce a larger cost. Since the MPSCP is $\mathcal{NP}$-hard, EX2 was only able to provide optimal solutions for the 20 and 50 nodes problems, taking about 2 s and 10 mins, respectively. Eleven of the 100 nodes instances were solved to optimality, taking up to 106 days. Other instances of this set occupied the EX2 algorithm even longer, in one instance more than five months, without reaching a feasible solution.

    We tried to solve all of the uniformly random 100 nodes problems to optimality, but CPlex simply takes too much time. However, the intermediate solutions found by EX2 can still be used as lower bounds, as they represent minimal power topologies, but they are not connected. EX2 is not suitable for clustered instances, though, as the intermediate solutions are even worse than the simple lower bound given by half the cost of the MST solution. The intermediate topologies tend to separate the clusters, thus leaving out the most expensive edges. We also refrained from trying to solve the larger problems. As the computation time of CPlex increases exponentially with the problem size, the benefit-cost-ratio is simply too low.

Results

Before presenting the results for the different local searches in the ILS, we want to evaluate how much can be gained compared to the simple MST approximation. Table 27 shows the best improvements found in any of the ILS runs, calculated as $1 - \text{Best}/\text{MST}$. As there are 30 instances in each problem set, the table shows the minimal, the average, and the maximal improvement for each problem set. The results show that the MST approximations can be improved by between 0.26 % and 19.28 %. For the unclustered instances, the average improvement seems quite constant at around 6.5 %. This observation was also made by Althaus *et al.* in [2, 3]. However, for some of these instances, we found improvements as high as 12.25 %.

    Clustered instances allow larger improvements. E. g. in the clustered 100 nodes instances (100c10), the average improvement is as high as

EX2

20 and 50 nodes

100 nodes

lower bounds

possible improvements

	Improvements to MST		
Size	min	avg	max
20	0.26 %	4.41 %	12.25 %
50	2.15 %	6.54 %	10.11 %
100	2.60 %	6.09 %	10.07 %
200	3.00 %	6.53 %	9.64 %
500	5.36 %	6.63 %	9.14 %
1000	5.21 %	6.58 %	7.62 %
100c10	1.48 %	10.58 %	19.28 %
200c14	2.11 %	6.13 %	12.26 %
500c22	2.40 %	5.26 %	11.81 %

Table 27: Improvements of best known solutions compared to the MST. For each problem set the minimal, average and maximal improvement is shown.

clustered instances

10.58 %, the best improvement is 19.28 %. However, when the number of clusters increases, the possible improvements quickly reduce to the aforementioned value of 6-7 %, due to closer or overlapping clusters. The higher improvements for clustered instances are a result of the links connecting different clusters. In the MST, these links connect the closest nodes from both clusters, and thus often force multiple nodes in the same cluster to have a higher transmission range. In the optimal solutions, only a few nodes or even only one node from each cluster is required to have a large transmission range to connect its cluster to other clusters. An example for this effect is shown in Figure 53. In a way, this is a similar effect as in the tightness example for the performance ratio of the MST in Figure 47, where small clusters with only two nodes have to be connected.

For the remaining results, we use the following table format: Again, each row gives the average values for all 30 problem instances of the

table format

same problem set. The first column denotes the problem size, where clustered instances are shown as 100c10, 200c14 and 500c22. For each of the considered heuristics, three columns show the calculation time in CPU seconds, the excess over the optimum or the best known solutions, and the number of runs out of all 900 runs that found these solutions. The best known solutions are taken from all experiments using

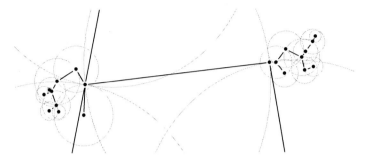

Figure 53: Detail from the best known solution for p100c10.0. From each of the shown clusters, only one node has a higher transmission range to reach another cluster. In the MST, at least two nodes from each cluster are chosen to build the inter-cluster connections.

the ILS. We believe only a few of these solutions up to $n = 500$ to be not optimal, but for $n = 1000$ the best known solutions can probably still be improved. An additional row gives the average results over all problem sizes for the respective heuristic to allow an easier comparison.

Table 28 shows results for the established local searches ES and EFS. These local searches were used to improve the MST solutions. Here, the ES already produces solutions that are less than 1 % above the optimum for the uniformly random networks. However, optimal solutions are found very rarely. Also, clustered instances pose a challenge for this simple heuristic. Using the stronger local search EFS reduces this gap. However, the high time complexity prohibits an application of this heuristic to larger instances. For the 1000 nodes problems we estimate running times of more than 100 hours, but already the running times for the 500 nodes problems are too high for practical use. The aim of our ILS heuristic is to produce results similar to EFS in less time.   *MST+ES*   *MST+EFS*

The actual results of the ILS using different local searches and mutations are shown in Table 29. Already the ILS using the weakest local search ST gives solutions that are better than the application of the simple ES heuristic to the MST. For the instances with up to 100 nodes, the results are even better than the EFS heuristic. In these experiments, we used the edge exchange mutation and also increased the number of iterations to 2000. As ST has a time complexity of only $\mathcal{O}(n^2)$, it can   *ILS*   *ST*

Size	ES			EFS		
	Time	Excess	Best	Time	Excess	Best
20	0.00 s	0.250 %	22	0.00 s	0.092 %	27
50	0.00 s	0.541 %	8	0.07 s	0.243 %	18
100	0.00 s	0.542 %	1	1.67 s	0.143 %	14
200	0.08 s	0.518 %	0	48.07 s	0.140 %	6
500	1.94 s	0.593 %	0	4 261.16 s	0.188 %	0
1000	23.71 s	0.549 %	0			
100c10	0.01 s	2.446 %	0	1.69 s	0.246 %	11
200c14	0.09 s	1.596 %	0	44.96 s	0.110 %	10
500c22	1.87 s	1.154 %	0	4 244.00 s	0.150 %	0
avg:	3.08 s	0.910 %	3	(1 075.20 s	0.164 %	10)

Table 28: Results for the simple heuristics ES and EFS starting from the MST

easily be applied to much larger instances than the ones considered here.

ES

However, better results can be found using the ILS with the stronger local search ES. In these experiments, we used only 200 iterations and kept the edge exchange mutation. Optimal solutions are found with very high probability in instances of up to 100 nodes. This heuristic can be applied to the larger set of 1000 nodes instances, but the running times are still unacceptably high.

filtered ES

The effect of the filters can be seen when comparing the results of the ILS using filtered and unfiltered ES. Using filters, the ILS takes only a third of the time and still produces better results for the larger instances. For the clustered instances (100c10 and 200c14) the average solution is worse than without filters, but the best known solutions are still found more frequently. The higher average excess is the result of some runs ending in poor local optima.

EFS

Using the strongest EFS local search in the ILS produces the best results. Optimal solutions can be found with very high probability. However, the time complexity of $\mathcal{O}(n^4)$ prohibits the application of this local search to larger problem instances. In our environment, $n = 200$ was the largest setting that delivered results in acceptable time. Using filters would allow to work on slightly larger problem instances, but the

Size	ST			ES (unfiltered)			ES (filtered)			VNS			EFS		
	Time	Excess	Best	Time	Excess	Best	Time	Excess	Best	Time	Excess	Best	Time	Excess	Best
20	0.05 s	0.046 %	849	0.03 s	0.000 %	900	0.01 s	0.000 %	900	0.04 s	0.000 %	900	0.11 s	0.000 %	900
50	0.30 s	0.073 %	663	0.49 s	0.000 %	899	0.15 s	0.005 %	871	0.90 s	0.002 %	886	4.64 s	0.000 %	900
100	1.25 s	0.123 %	396	4.86 s	0.016 %	748	1.44 s	0.016 %	797	11.64 s	0.004 %	869	88.55 s	0.002 %	891
200	5.39 s	0.210 %	66	50.34 s	0.118 %	45	14.39 s	0.071 %	216	174.91 s	0.006 %	770	1 909.85 s	0.003 %	846
500	44.08 s	0.517 %	0	1 291.41 s	0.330 %	0	342.88 s	0.201 %	1	8 884.07 s	0.028 %	271	130 516.92 s	0.016 %	371
1000	238.68 s	0.770 %	0	16 452.88 s	0.523 %	0	4 421.94 s	0.377 %	0	214 577.03 s	0.053 %	30			
100c10	1.07 s	0.120 %	380	5.06 s	0.085 %	413	2.27 s	0.222 %	433	27.84 s	0.004 %	814	200.24 s	0.001 %	881
200c14	4.72 s	0.191 %	47	50.43 s	0.176 %	47	21.96 s	0.196 %	162	519.27 s	0.016 %	738	4 529.82 s	0.001 %	847
500c22	41.08 s	0.331 %	0	1 267.00 s	0.234 %	0	384.31 s	0.206 %	0	17 548.23 s	0.026 %	131	213 448.83 s	0.009 %	221
avg:	37.40 s	0.265 %	266	2 124.72 s	0.165 %	339	576.59 s	0.144 %	375	26 860.44 s	0.015 %	601	(43 837.37 s	0.004 %	732)

Table 29: Results of the ILS for the MPSCP. Average results for all instances of the same size are shown for different local searches. The local searches are sorted from weak to strong: Subtree Moving Search (ST), Edge Switching Search (ES), without filters and with filters, Variable Neighbourhood Search (VNS), and Edge and Fork Switching Search (EFS). For each local search, the CPU time, the excess above the best known or optimal solution, and the number of runs that found this solution is shown. As there are 30 instances for each size and the experiments were repeated 30 times, there are 900 runs for each setting.

time complexity remains the same. We used the ILS with EFS only for finding the best known solutions, and deactivated the filters in order to get the best results.

VNS

The VNS combines the advantages of ES and EFS. It produces only slightly worse results than the full EFS but takes only a fraction of the CPU time. Comparing the results against the simple ES shows that VNS reduces the gap by a factor of 10. This local search was also used to find the best known solutions for the 1000 nodes problems, accepting the running times of almost three days for each instance of this size.

comparison

The best heuristic in terms of running time and solution quality is the ILS using ES local search together with random increase mutation and filters. The fastest heuristic is the ILS with ST local search, and best results are found using ILS with the strongest local search EFS, ignoring the high calculation times, or with the VNS which reduces the running times. For all considered local searches, the running times are still much lower than the calculation times of the exact solver EX2 from [123].

5.4.4   Discussion

In this section, we have presented an Iterated Local Search (ILS) for the MPSCP. We evaluated three different local searches: ES, EFS and a Sub-tree Moving Search (ST). A similar version of ST was already success-

ILS

fully applied to the MRCST. A filtering technique proposed by Montemanni *et al.* in [123] was integrated in the ILS, resulting in smaller calculation times as well as smaller gaps to the optimal solutions, as it helps avoiding unnecessary calculations. This ILS using a slower implementation of the EFS based on deletion search has already been published in [183]. In this thesis, we used an insert search version of the EFS, which produces the same results, but is faster than the deletion search version. We also introduced a simple but powerful VNS that combines the established local searches ES and EFS.

results

The ILS was tested on nine sets of problem instances of various sizes and configurations. The problem sets contained uniformly random instances with up to 1000 nodes as well as clustered instances. The experiments showed that the ILS is able to find near-optimal solutions in short time. We also used the ILS with the strong VNS to find best known solutions for the largest problem set, accepting the higher running times. This allowed us to establish a base line for comparison for the distributed algorithm presented in the following section.

Obtaining optimal solutions using exact algorithms proved to be very expensive. Using our own implementation of the exact algorithm EX2 from [123], we were able to prove the optimality of the best known solutions for all problem instances with 20 or 50 nodes. Computation times of a couple of seconds or minutes are usual for these sizes. As the MPSCP is $\mathcal{NP}$-hard, the computation times of EX2 increase exponentially. Only eleven of the 100 nodes instances were solved this way. As the computation times already increased to a couple of months for this problem size, we were unable to prove the optimality of the remaining instances.

proof of optimality

5.5 DISTRIBUTED RANGE ASSIGNMENT PROTOCOL

In this section, we demonstrate how the wireless network can be used as a self-optimizing system to reduce the energy needed for data transmissions. We assume that the wireless nodes can adjust their transmission power and that a higher transmission power corresponds to a larger coverage area for this node. We present a fully distributed algorithm for creating and maintaining power-efficient topologies in a wireless network. The algorithm is designed to first create a connected topology and then iteratively search for better connections to reduce the overall power consumption.

self-optimizing wireless network

This section is structured as follows: Section 5.5.1 gives an overview of related work concerning distributed algorithms for the MPSCP. In Section 5.5.2 we present our distributed algorithm. Section 5.5.3 gives results of simulated experiments on the same sets of problem instances that were used for the global optimization. This section is concluded in Section 5.5.4.

structure

5.5.1 Related Work

Some of the publications for the global view optimization shown in Section 5.4.1 already indicate or explicitly suggest distributed versions of their algorithms. An example is the IPP presented by Cheng *et al.* in [25]. This heuristic builds the tree in a way that resembles Prim's algorithm for building MSTs [145], but it needs global knowledge and can only be transformed to a distributed algorithm with a high level of cooperation between the nodes.

distributed IPP?

distributed ES?

Althaus *et al.* presented the ES local search in [2, 3]. They also tried to limit the number of hops for the ES heuristic, as this translates to the level of cooperation in a distributed algorithm. However, they do not proceed in this direction and do not present a fully operational distributed algorithm.

no fully distributed algorithm so far

No fully developed distributed algorithm for the MPSCP has been proposed so far. There are distributed algorithms for the MST which could be used to construct a 2-approximation to the MPSCP. But the challenge of designing a distributed algorithm for the local search heuristics that includes propagation of routing information and cycle detection, and that works without a high level of cooperation, has not yet been tackled.

distributed MEB

There are, however, a number of proposals for distributed algorithms for the MEB: Wieselthier *et al.* already proposed a distributed version of their BIP in [179]. The results are only slightly worse than BIP. Bin Wang and Sandeep K. S. Gupta present S-REMiT in [176]. In this algorithm, each node tries to improve the tree by connecting to another parent. The presented algorithm is a distributed algorithm, however, it uses a token to allow only one node at any time to apply a change. Xiang-Yang Li *et al.* present a fully distributed algorithm for the MEB in [100] that constructs a locally structured network by building a k-local MST. Sung Guo and Oliver Yang present a distributed algorithm for multicasts in wireless networks in [62], that also copes with mobility.

argument against distributed MEB

The main problem when designing distributed algorithms for the MEB is that the algorithm needs to receive information from the leaf nodes of the broadcast tree, while the MEB theoretically allows these nodes to switch off their transmission unit completely. Also, when used in real world scenarios where transmissions could be lost due to interferences, a reliable broadcast can only be achieved by having the nodes acknowledge the successful reception of the broadcast message. As both issues can be solved by creating bi-directional links, we argue that a strict application of the MEB topologies is not necessarily the best approach. Instead, the topologies created by the MPSCP could also be used for broadcasting.

5.5.2   Distributed Range Assignment Protocol

The distributed algorithm DistRAP presented in this section is supposed to be run on every node in the wireless network. Each node establishes and dissolves connections to other nodes according to the rules of the algorithm. The algorithm uses messages to find expensive connections that can be removed. The nodes also probe for new links in order to find less expensive connections.

In DistRAP, each node maintains a list of neighbours, a list of links, and a routing table. The term neighbours refers to all nodes that can be reached using the maximum transmission range of the node. Only a subset of these neighbours is used for the actual topology. A DistRAP node establishes and revokes links to other nodes by sending control messages. These links form the actual topology. The longest link determines the transmission range of the node and corresponds to the cost of this node in the MPSCP. Each DistRAP node becomes active from time to time to try and establish new links to improve the current topology.   *neighbours and links*

As the constructed wireless topology forms a multi-hop network, nodes are required to relay messages for other nodes and react on incoming control messages. They carry a routing table to be able to pick the corresponding link to reach the target node. These routing tables could be avoided, but this would create the necessity of flooding large parts of the network for every single message.   *routing table*

When a node is activated, it sends out a broadcast PING message to announce its presence to its neighbouring nodes. Other nodes that receive this signal add the new node to their neighbour list and respond with a PONG message. The signal strength of this response messages is used to determine the transmission power necessary for establishing the link.   *joining*

The DistRAP algorithm works in two phases: In the first phase, a simple but probably not energy minimal topology is constructed to connect all nodes of the wireless network. In the second phase, this topology is improved step by step. However, as there is no central entity announcing the end of a phase and the beginning of the next phase, some of the wireless nodes can already reach phase two while other nodes are still in phase one. This means that some parts of the network are already being improved while other parts of the network are not yet fully connected.   *two phases*

In the first phase of DistRAP, the wireless node actively sends out ADDLINK messages to the closest not yet reached neighbour. This neighbour is determined by the neighbour list and the current routing table. This neighbouring node also checks its routing table and if it does not know another route to the initiator node, it responds with a CONFIRM-   *phase one*

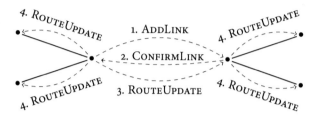

Figure 54: Message flow in DistRAP for adding a new link. The initiating node
sends an ADDLINK request to a not yet reached node, which con-
firms this request if it also does not know a route to the initiating
node. After the new link is established, both nodes send ROUTEUP-
DATE messages to their part of the tree to make the new link known.

LINK message. This establishes the link. The final step is an update
of the routing tables. To this end, the CONFIRMLINK message already
contains the current routing table of the neighbouring node, which is
merged with the routing table of the initiator node. The updated routing table is then propagated to all connected nodes. This process of
adding a new link is shown in Figure 54.

In the second phase of DistRAP, the node intentionally send an ADD-
LINK message to a neighbouring node that already is contained in the
phase two   routing table. As inserting this link would introduce a cycle, the neighbouring node sends a SCOUT message back to the origin node using the
routing information. This SCOUT is relayed along the cycle and each
node adds information on how much energy it could save if it were to
remove the links the SCOUT is taking. The savings for each link need to
be calculated by two nodes, as two nodes are connected through this
link and could possibly reduce their transmission ranges when the link
distributed   is removed. The SCOUT message is also used for this calculation: The
calculation of   first node stores its savings in the SCOUT message, the second node
savings   adds its own savings to this value and updates the information about
the best savings, which are also stored in the SCOUT. When the SCOUT
returns to the origin node, this node finishes the calculation for the
last link of the cycle and checks whether the best savings indicate that
the new link should be removed or kept. If removing another link reduces the total cost, the new link is accepted. The origin node sends
DELETELINK messages to the nodes adjoining the link that needs to
be removed. An ADDLINK message is sent again to the neighbouring

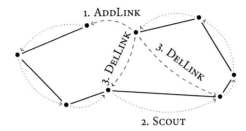

Figure 55: Message flow in DistRAP for replacing a link. The initiating node sends an ADDLINK request to an already reachable node, which sends a SCOUT along the induced cycle back to the original node. This SCOUT message collects information on the energy needed for the links of the cycle. The link that needs the most energy is removed from the cycle and the new link is established if this improves the topology.

node to establish the new link. Using the new routing information, the neighbouring node should now respond with a CONFIRMLINK message establishing the link. This process is shown in Figure 55. This phase of the protocol essentially represents a distributed version of the ES local search.

As DistRAP is designed to create a self-organizing wireless network, it does not enforce any synchronization upon the nodes. Due to the time needed for propagating the routing updates, nodes may operate on outdated information and establish an unnecessary link. This link also creates a cycle in the topology, but the corresponding nodes will not recognize their mistake until they receive the routing update. As the routing information for this new link is also propagated to the network, some remote nodes will detect the cycle, because they receive contradicting routing information. Whenever a node detects the cycle, a cycle repair mechanism is started.

unwanted cycles

The simple solution of cutting a random link from the cycle does not yield useful results in a distributed environment, as the cycle is often detected by a large number of nodes and a large number of links of the cycle would be cut, resulting in a solution far worse than before. DistRAP takes another approach. Similar to the intentionally induced cycles in phase two, the nodes determine the link to be removed from the cycle based on the best savings. One minor obstacle is the lack of

find best link to be removed

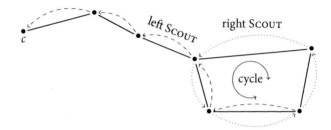

Figure 56: Message flow in DistRAP for detection and resolving a cycle. If a
node receives contradicting routing information for another node c,
it sends out two SCOUT messages to c, one for each direction. The
SCOUT is rerouted back to the initiator node, if a node knows another
route. The SCOUT that actually reaches node c is ignored. The right
SCOUT, i. e. the SCOUT that traversed all links of the cycle, contains
information on which edge from the cycle should be removed.

a global view. The nodes participating in the cycle do not know the
remaining nodes from the cycle, even after they recognized the cycle.
They only know a subset of nodes for which they have received contradicting routing information. From this information, they can determine only two links from the cycle. In the cycle repair mechanism
of DistRAP they simply pick a node c with contradicting routing information. One routing entry points in one direction of the cycle and
another routing entry points in the opposite direction of the cycle. As
can be seen in Figure 56, this node c is not necessarily part of the cycle,
it can also be several hops away from a node of the cycle. The algorithm has to take care that only links of the actual cycle are considered
to be removed. The path from the cycle to node c must not be severed,
otherwise the network would be split and the cycle would remain.

cycle repair
mechanism
In DistRAP, the node that recognizes the cycle sends two SCOUT messages to node c, one message for each route. If the SCOUT passes a node
that knows a different route back to the origin node, it sends the SCOUT
back over this route. If the SCOUT reaches the node c, but this node c
does not know a different route back, the SCOUT is ignored. Only one
SCOUT will return to the origin node after traversing only links of the
cycle, and the origin node can decide which link should be removed
and send the necessary DELETELINK messages.

5.5.3 Experiments

The experiments shown in this section are based on a simulation frame- *simulation*
work written in C++. Each node was simulated as a separate thread   *framework*
resulting in a high amount of concurrency. The concurrency was increased again by conducting these simulations on a cluster of Dual
CPU Quad Core 2.5 GHz Intel Xeon machines. In essence, eight node
threads were processed at the same time. The simulation framework
allowed nodes to transmit and receive messages. Received messages
were queued, such that the nodes could process them as soon as they
acquired a CPU time slot. The simulation did not cover interference
or lost messages. It also did not distinguish between simulated time
and real time. To simulate the time for transmitting or receiving a message, the simulator introduced a 50 ms wait before a node was allowed
to process an already received message.

The wireless nodes became active after random waiting intervals be- *settings*
tween 1 and 10 seconds. In phase one, the nodes establish links to their
neighbours. As the routing updates need to propagate through the network, the nodes wait a similar random interval before trying to add
a new link. In phase two, when the nodes actively induce cycles, the
same random waiting interval is used. So on average, each node tries
to add one link to the topology every 5.5 seconds.

For our experiments, we used the same sets of test instances as in
Section 5.3.5 and Section 5.4.3. Each set contains 30 instances of the *instances*
same size. Six of these sets contain uniformly random instances, while
the remaining three sets contain clustered instances. Each simulation
was repeated 30 times and average values are used for the following
discussion.

Results

As DistRAP improves the wireless topology over time, we use plots to
present the results. In these plots the average excess over the optimal or
best known topology from Section 5.4.3 is plotted over time, averaged
over all 30 runs for each of the 30 instances of the same size. An excess
of 0 % means that an optimal topology has been found.

Figure 57 shows the results for the uniformly distributed networks.
During the first 10 seconds the nodes are activated. Once they have *uniformly*
found neighbours, they enter phase one and try to establish links. Dur- *random*
ing this phase, which lasts about 20 seconds, the overall transmission *networks*

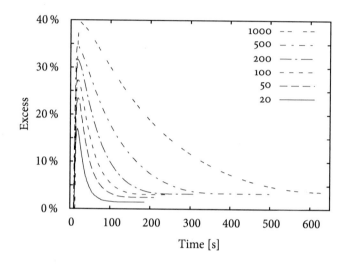

Figure 57: Results for uniformly distributed random networks

energy rises as the wireless nodes increase their transmission ranges to accommodate the newly established links. When the network is connected, the excess above the optimal solutions is still very high. Values between 17 % and 40 % are usual. By this time, some nodes have already entered phase two and started replacing the most expensive links. This phase lasts longer than phase one, as each improvement has to be confirmed by all nodes of the cycle. Also, the improvements become smaller as the topology approaches the optimal solutions. The results of this topology change can clearly be seen in Figure 57. The figure also shows that the transformation is slower for larger networks, as the number of possible links as well as the average cycle length increases. The final excess over the best known topologies lies between 1.4 % and 3.6 %.

Another noteworthy effect concerns phase one, during which the initial topology is constructed. DistRAP nodes try to add smaller links *slow first, then* before larger links. Thus, the total cost starts to increase slowly in the *faster* beginning, but increases faster toward the end of phase one. Unfortunately, this effect cannot be seen in the figures, as the corresponding curves would have to be drawn between -100 % and 0 % excess.

clustered The results for the clustered networks are shown in Figure 58. Here, *networks* the initial excess as well as the final excess is higher than in the uni-

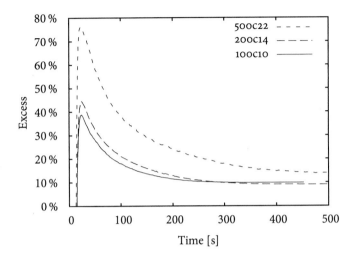

Figure 58: Results for clustered random networks

formly distributed networks The initial excess lies between 39 % and 76 %, and the final excess between 9 % and 14 %. The higher excess can be explained by the inter-cluster links. In clustered instances, often only one node from each cluster maintains connections to other clusters (see Figure 53 on page 201). A node that establishes two or more of these longer connections during phase one is in the unfortunate position to maintain these connections. As DistRAP is based on the ES local search, it cannot transfer two links at the same time to another node. However, the amount of energy that actually is saved in the clustered networks is higher than in the uniformly random networks.

5.5.4 Discussion

We have presented a self-organizing distributed algorithm to reduce the necessary transmission energy in wireless networks. In simulated experiments on a wide range of test instances with up to 1000 nodes, both uniformly distributed and clustered, we have shown that the algorithm can reach near-optimal topologies. Preliminary results of this algorithm have already been published together with Tom Ansay in [185].

DistRAP

In the wireless setting, the saved energy translates to a longer network lifespan without increasing the batteries of the nodes. As a side effect, the amount of interference in the wireless network is also reduced due to the reduced transmission ranges.

MEB

The topologies constructed by DistRAP can also be used for broadcasts. However, we refrained from optimizing the Minimum Energy Broadcast (MEB) objective directly, as it does not allow the leaves of the broadcast tree to acknowledge the reception of the broadcast. In a real wireless network, this acknowledgement is vital, as wireless communication can easily be interrupted by interference, movement of the nodes or untimely sleep states. If the broadcasting node depends on this acknowledgement, then the Minimum Power Symmetric Connectivity Problem (MPSCP) is a better representation for the total objective.

5.6 SUMMARY

This chapter has shown two problems of saving energy in wireless networks. The problems assumed different communication schemes. The Minimum Energy Broadcast (MEB) can be used to reduce the necessary transmission energy for a broadcast, while the Minimum Power Symmetric Connectivity Problem (MPSCP) can be used to create symmetrically connected topologies. We have shown that the nodes of the wireless network can be used to improve the topology of the network in this respect. The distributed algorithm DistRAP was shown to considerably reduce the amount of energy needed to connect the network.

self-optimization

The results of the distributed algorithm have been compared to optimal solutions found by an exact solver EX2 by Montemanni *et al.* [123]. As the MPSCP is an $\mathcal{NP}$-hard problem, EX2 already took days proving the optimality of the best known solutions for the unclustered 100 nodes problems. For all larger instances as well as the clustered instances, we used the best known results from global view optimization as a base for comparison.

optimal solutions

For the global view heuristics, we followed several paths. We developed several heuristics based on local search. Some of the used local searches were already established, such as the Edge Switching Search (ES), the Edge and Fork Switching Search (EFS), or the r-shrink local search. Other local searches have been introduced in this thesis, such as the Variable Neighbourhood Search (VNS), or in our publications, such as the Subtree Moving Search (ST) and modifications of the r-shrink local search. As both the MEB and the MPSCP produce tree

global view heuristics

topologies, we used a set of simple tree operations for the mutation in   *ILS and ELS*
our Iterated Local Search (ILS) and Evolutionary Local Search (ELS).
We also presented a very different type of global view heuristic, the
Nested Partitioning (NP). The NP builds the solution by finding and   *NP*
collecting good solution parts, the arcs in the wireless topology. The
original NP algorithm already produced good results, but they were
further improved by introducing a simpler lower bound, and more importantly by integrating an Ant Colony Optimization (ACO) approach.
The ACO was also used without the NP corset and produced quite re-   *ACO*
markable results.

CONCLUSION

6

In this thesis, we looked at a range of network optimization problems and presented several distributed algorithms as well as global view heuristics for solving them. While the distributed algorithms modelled these networks as a self-organizing network where each node contributes to the global optimization goal by applying simple local rules, the global view heuristics considered the network as a whole and used knowledge about all parts of the network during the optimization.

global and distributed

The considered optimization problems are not just restricted to computer networks. They also appear in areas such as airplane traffic or mail transport, or other forms of communication networks. However, computer networks are a perfect choice for implementing distributed algorithms. The nodes in these networks are already equipped with computational resources as well as means for communication or storing data.

self-optimizing

The concerned networks can be split in two categories: wireless networks and wired computer networks. However, the global view optimization works on a high level model and abstracts from these physical settings. Many of the techniques from the wired networks were successfully transferred to the wireless networks.

graph representation

The global view optimization heuristics used in this thesis are based on local search. We used established local searches (r-shrink, Largest Expanding Sweep Search (LESS), Edge Switching Search (ES) and Edge and Fork Switching Search (EFS)), presented our own local searches (Subtree Moving Search (ST), replace, swap and reassign for the SPSP, a Variable Neighbourhood Search (VNS) combining ES and EFS for the MPSCP) or transferred local searches from other problems (Ahuja-Murty Local Search (AMLS) from the OCST).

local search

Most of the proposed global view heuristics are based on some form of Memetic Algorithm (MA), such as the Evolutionary Local Search (ELS) and the Iterated Local Search (ILS). In the ELS, the population size is reduced to one individual and no recombination is used. As the ELS uses strong local searches, this was already sufficient to reach optimal or near-optimal solutions. Using larger populations would improve the average solutions for the larger problems, but it would also

ELS and ILS

lead to larger computation times. The ILS approach follows the opposite idea: By reducing the number of offspring while using even stronger local searches, the same solution quality and computation times can be achieved as with an ELS using a weaker local search.

mutation

Both approaches need some form of mutation to escape from the local optima of the used local searches. We used a wide range of mutation operators (edge exchange, random range increase, random swaps, random subtree moves) as well as a general mutation rate adaptation scheme that allows the heuristics to adapt to the best mutation rate for the individual problem instance as well as the current search progress. Using this mutation rate adaptation, our heuristics are able to explore the search space during in the beginning of the search and concentrate on good parts of the solution space during the later stages of the search, thus exploiting the structure of the best solutions found so far.

NP and ACO

We also used other global view heuristics, such as the Nested Partitioning (NP) or Ant Colony Optimization (ACO) approaches, or combinations of these. NP is a very recent heuristic. It works by iteratively partitioning the search space and constructing a solution by selecting good solution components piece by piece. The selection is guided by a random sampling procedure, which can also be replaced by other heuristic methods, such as the ACO. ACOs, when used in combination with a local search, can be seen as a replacement for the mutation operator of an ELS. They work by repeatedly constructing new solutions, reusing information from all previous runs. This information is stored in form of pheromones placed on the solution components. Those components that lead to good solutions receive more pheromones than other solution components. The construction scheme used in ACOs uses this information as a probability value and randomly constructs new solutions. Although these solutions are newly generated, they can be viewed as a form of mutation of previous solutions. While the ELS and ILS algorithms used a completely random mutation that did not value the relevance of the changed solution components, the ACO approach allows to confine the search to the most promising parts of the solution space.

optimal results

The results of the global view heuristics were compared against optimal solutions found by Mixed Integer Programming (MIP) or similar approaches. These MIP formulations were taken from the literature, however, we improved one formulation and also transferred tree formulations for one problem to other problems. This comparison showed that the global view heuristics used in this thesis produced close-to-optimal solutions in relatively short time. As we wanted to use the best known solutions as a base line for comparison for the distributed algorithms, we also used excessively high parameters for the global view heuristics to improve these solutions, accepting the high running times.

For the distributed algorithms we proposed the following path: We transferred parts of the global view heuristics to distributed algorithms, whenever this was possible. As the level of necessary cooperation in a self-organizing system should be kept low, this approach was not always practicable. In these cases, we proposed distributed algorithms that solve an alternate problem, such as the Shortest Path Tree (SPT) in the case of the Minimum Routing Cost Spanning Tree (MRCST). *distribution*

In the case of the Super-Peer Selection Problem (SPSP), even the local view of the distributed algorithms was too large. Determining all values for adjacent links meant putting an unacceptably high load on the network. We successfully used estimation techniques to reduce this load. Network coordinates proved to be an effective technique to reduce the amount of Round Trip Time (RTT) measurements, while still giving adequate estimations. *estimations*

We also proposed to further improve the created network topologies by changing parts of the original optimization problem. This was done in the SPSP, where the distributed algorithm ChordSPSA replaced the fully-meshed intra-core connections by a Chord ring. In such cases, the original mathematical formulation as well as the results from the global view optimization could no longer be used to find optimal solutions. Instead, we used simpler lower bounds. We also compared the constructed topologies to unoptimized topologies. This shows the gain that is achievable by optimization. *change the optimization problem*

The proposed algorithms have been simulated, and the results were discussed in this thesis. The simulation covered major aspects of self-organization, as the nodes worked on a local view and transferred messages in order to apply changes or to inform other nodes about such changes. We used a simulation framework that allowed to re-use the code in a real distributed environment. Although this work is not part of this thesis, some of these algorithms have already been successfully deployed in real world experiments and are part of further research. One example are the delay-optimized Peer-to-Peer (P2P) topologies created by ChordSPSA which are used in a desktop grid called Peer-Grid developed by Matthias Priebe [144]. *simulation* *application results*

The distributed algorithms presented in this thesis can be used to create self-organizing networks that optimize their own topology in order to reduce the communication cost. Whether this cost is given by the message delay or the necessary transmission power, applications using such networks benefit largely from the optimized topologies.

APPENDIX

In this appendix, additional information on the considered optimization problems and more detailed result tables for the proposed heuristics are provided. The appendix follows the structure of the thesis, as each section corresponds to one chapter, describing one optimization problem.

A.1 MINIMUM ROUTING COST SPANNING TREES

A.1.1 *Fitness Distance Analysis*

Table 30 gives detailed information on the local optima for all considered local searches for the Minimum Routing Cost Spanning Tree (MRCST) and all considered PlanetLab instances. The minimal, maximal, and average values from this table were used in Table 6 on page 56.

FDC analysis

A.2 SUPER-PEER SELECTION PROBLEM

A.2.1 *$\mathcal{NP}$-hardness of the SPSP*

The Super-Peer Selection Problem (SPSP) may be cast as a special case of the Uncapacitated Single Assignment p-Hub Median Problem (USA-pHMP) [131, 133, 51, 166], which is known to be $\mathcal{NP}$-hard. In both problems, p nodes are to be selected out of n nodes to serve as hubs (or super-peers). The remaining nodes are assigned to one hub, respectively. Hubs are assumed to be assigned to themselves. Denoting the assignment of node i to hub j by a binary variable $x_{ij} = 1$, the transportation costs for the link between i and j by d_{ij}, and the amount of flow from i to j by w_{ij}, the cost of such a hub location can be defined as

special USApHMP

$$C(X) = \sum_{i=1}^{n}\sum_{j=1}^{n}\sum_{k=1}^{n}\sum_{l=1}^{n}\left(d_{ik} + d_{kl} + d_{lj}\right) \cdot w_{ij} \cdot x_{ik} \cdot x_{jl}. \qquad \text{(A.1)}$$

LS	STconst			STrand			CYCrand			ST2opt			CYC2opt			AM		
	Exc	FDC	Dist	Exc	FDC	Dist	Exc	FDC	Dist	Exc	FDC	Dist	Exc	FDC	Dist	Exc	FDC	Dist
01-2005	45.7	0.26	78	10.4	0.48	65	10.8	0.60	81	2.4	0.57	39	1.8	0.61	34	1.1	0.38	40
02-2005	35.0	0.10	193	10.0	0.56	143	9.7	0.69	193	2.1	0.60	61	1.5	0.57	64	1.3	0.66	67
03-2005	32.9	0.11	203	9.9	0.52	155	8.5	0.69	189	2.2	0.65	75	1.5	0.78	74	1.1	0.69	57
04-2005	14.3	0.43	35	5.5	0.48	28	7.5	0.77	42	0.9	0.40	17	0.4	0.72	12	0.5	0.77	13
05-2005	69.4	0.05	237	29.2	0.14	174	25.1	0.20	230	1.8	0.44	56	1.4	0.65	55	0.8	0.87	47
06-2005	38.3	0.09	227	12.8	0.60	164	10.9	0.61	211	2.2	0.76	63	0.8	0.74	46	1.6	0.79	59
07-2005	38.3	0.10	230	17.6	0.10	162	16.2	0.14	223	1.8	0.62	64	1.2	0.74	55	1.5	0.79	69
08-2005	24.1	0.17	222	6.5	0.72	153	6.8	0.74	231	1.4	0.81	50	0.8	0.83	38	1.3	0.86	55
09-2005	58.2	0.08	248	27.2	0.16	180	23.4	0.23	240	2.3	0.48	63	0.4	0.53	41	1.0	0.48	52
10-2005	30.5	0.13	225	8.1	0.65	148	8.3	0.69	230	1.6	0.87	46	1.1	0.92	33	0.4	0.93	19
11-2005	31.4	0.15	222	8.4	0.62	154	7.8	0.75	234	1.6	0.84	50	0.9	0.85	36	0.5	0.86	16
12-2005	60.1	0.09	246	24.5	0.13	176	20.7	0.23	229	1.6	0.56	58	1.1	0.71	55	0.7	0.64	54
average:	39.9	0.15	197	14.2	0.43	142	13.0	0.53	194	1.8	0.63	53	1.1	0.72	45	1.0	0.73	46

Table 30: Detailed properties of the local optima. For each local search neighbourhood and for each considered PlanetLab instance, the average excess over the best known solution (Exc), the Fitness Distance Correlation coefficient (FDC), and the average distance to the best known solution (Dist) are shown.

This cost is to be minimized. There are a number of constraints:

$$x_{ij} \leq x_{jj} \qquad\qquad i, j = 1, \ldots, n \qquad\qquad (\text{A.2})$$

$$\sum_{j=1}^{n} x_{ij} = 1 \qquad\qquad i = 1, \ldots, n \qquad\qquad (\text{A.3})$$

$$\sum_{j=1}^{n} x_{jj} = p \qquad\qquad\qquad\qquad (\text{A.4})$$

$$x_{ij} \in \{0, 1\} \qquad\qquad i, j = 1, \ldots, n \qquad\qquad (\text{A.5})$$

The set of constraints (A.2) ensures that nodes are assigned only to hubs, while (A.3) enforces the allocation of a node to exactly one hub. Due to constraint (A.4), there will be exactly p hubs. Now, the SPSP is the special case of the USApHMP, where all flow between different nodes $i \neq j$ is $w_{ij} = 1$ and no flow from a node to itself is generated: $w_{ii} = 0$.

The general USApHMP is known to be $\mathcal{NP}$-hard. In fact, even if the *general* set of hubs is fixed, the remaining assignment problem is still $\mathcal{NP}$-hard, *USApHMP* as soon as $p \geq 3$ [166], but polynomial time algorithms exist for $p = 2$ or $p = 1$ [165]. The proof for $\mathcal{NP}$-hardness in [166] reduces the $\mathcal{NP}$-hard 3-Terminal Cut to USApHMP. The proof needs the weights to be arbitrarily selectable from $w_{ij} \in \{0, 1\}$. Since SPSP fixes the weights to $w_{ij} = 1$, the proof cannot be reused, and it is still unclear whether SPSP is $\mathcal{NP}$-Hard.

In the following, we provide a proof that the SPSP is $\mathcal{NP}$-hard. We show this by reducing the Max Clique problem [57, problem GT19] to SPSP. A slightly different version of this proof has already been pub- *reduction from* lished in [180]. *Max Clique*

Max Clique: Given a graph $G = (V, E)$ and a number $k > 0$, does there exist a subset $V' \subseteq V$ with cardinality $|V'| \geq k$, such that every two vertices in $i, j \in V'$ are joined by an edge $(i, j) \in E$?

Max Clique is $\mathcal{NP}$-hard [57]. Max Clique can be transformed to an SPSP by introducing a new vertex $a \notin V$, setting $W := V \cup \{a\}$, and setting the distance function d_{ij} to:

$$d_{ij} = \begin{cases} 0 & \text{if } i = a \vee j = a \vee i = j \\ 1 & \text{else, if } (i, j) \in E \\ 2 & \text{else} \end{cases} \qquad\qquad (\text{A.6})$$

THEOREM A.2.1. The decision problem whether there is a super-peer selection X selecting $p = k + 1$ hubs and resulting in a cost $C(X) \leq Z =$

$(p-1)\cdot(p-2)$ in the weighted graph $G' = (W, W \times W, d)$ is equivalent to the Max Clique decision problem in the original unweighted graph $G = (V, E)$.

Proof. We first show that a solution for the Max Clique Problem is also a solution for the SPSP. We then show that every solution for the SPSP is a solution for Max Clique.

Consider a solution for the Max Clique problem, i. e. a subset $V' \subseteq V$, $|V'| = k$, such that every two vertices $i, j \in V'$ are joined by an edge $(i, j) \in E$. The super-peer selection $X_a(V')$ that chooses all $p = k + 1$ nodes from the set $V' \cup \{a\}$ to be hubs, and assigns the remaining nodes to hub a, only creates costs on the inter-hub links. Since all distances to and from node a are zero, its links to the non-hub nodes and the remaining $p - 1 = k$ hub nodes again adds nothing to the total cost. The cost for this super-peer selection $X_a(V')$ is simply the sum of the remaining inter-hub links between the $p - 1$ hubs V':

$$C(X_a(V')) = \sum_{k \in V'} \sum_{l \in V'} d_{kl} \cdot x_{kk} \cdot x_{ll} = \sum_{k \in V'} \sum_{l \in V' \smallsetminus \{k\}} 1 = (p-1)\cdot(p-2) \leq Z$$

(A.7)

Thus, every solution for the Max Clique Problem is a solution for the SPSP.

Consider an arbitrary assignment X_b that selects p hubs and assigns the non-hub nodes to hubs. We will consider the following non-disjoint cases, and show that in each case the assignment X_b does not give a solution to the SPSP, i. e. its cost is larger than Z:

1. Node a is not selected as a hub in X_b.

2. A non-hub node is not assigned to a in X_b.

3. A set of nodes V' is selected as the hub-set in X_b, such that two of them $i, j \in V'$ are not joined by an edge $(i, j) \in E$.

If node a is not selected as a hub, all inter-hub links between the p hubs will have distances $d_{ij} \geq 1$, creating a cost of $C(X_b) \geq p \cdot (p-1) > Z$. Thus, every solution to the SPSP must select a as a hub. Since all other distances are $d_{ij} \geq 1$, the remaining inter-hub links create a cost of $H \geq (p-1) \cdot (p-2) = Z$.

If a non-hub node i is not assigned to a, its link to the hub $j \neq a$ creates additional cost of at least $d_{ij} \geq 1$. The total cost is then $C(X_b) \geq$

a not hub

*node not
assigned to a*

$H + d_{ij} > Z$. Thus, every solution to the SPSP must assign all non-hub nodes to hub a.

If two of the selected hubs $i, j \in V'$ are not joined by an edge $(i, j) \in E$, the corresponding distance is $d_{ij} = 2$. This increases the cost for the inter-hub links and thus the total cost to $C(X_b) \geq H + 1 > Z$.

unconnected hubs

In all cases 1, 2 and 3, the assignment X_b is not a solution to the SPSP, thus, every solution to the SPSP must select node a as a hub, assign all non-hub nodes to a, and select a fully connected set of nodes V' as hubs, which is also a solution to the Max Clique problem.   □

Note, that if X_b selects a different set of fully connected nodes as hub-set than X_a, it only constitutes a solution to the SPSP if node a is also selected as a hub and all non-hub nodes are assigned to a. However, the chosen set of hubs except node a is again a solution to the Max Clique problem.

A.2.2 *Parameter Study for the ELS*

Tables 31 and 32 give average results for the Evolutionary Local Search (ELS) with Don't Look Markers (DLMs) and different parameters κ and λ on the larger instances of the AP set ($n \geq 100$). The first table shows the average excess over the best known solutions, averaged over all runs and all instances. The second table shows the corresponding average CPU times.

with DLMs

Tables 33 and 34 show the results for the ELS without DLMs. The first table shows the average excess over the best known solutions, averaged over all runs and all instances. The second table shows the corresponding average CPU times.

without DLMs

A comparison between the ELS with or without DLMs shows that the markers speed up the ELS by a factor of about 2, but also decrease the solution quality. In both settings, the results with $\kappa = 20$ iterations are better than the results for $\kappa = 50$ iterations and $1/2 \cdot \lambda$ offspring, or $\kappa = 10$ iterations and $2 \cdot \lambda$ offspring.

Detailed results for $\kappa = 20$ and $\lambda = 100$ have been shown in Tables 18 and 19 on page 106.

	Excess over best known solution [%]						
λ	$\kappa = 10$	$\kappa = 20$	$\kappa = 50$	$\kappa = 100$	$\kappa = 200$	$\kappa = 500$	$\kappa = 1000$
10	1.84	1.26	1.13	1.07	1.02	0.95	0.89
20	1.40	0.95	0.84	0.81	0.76	0.70	0.65
50	0.98	0.62	0.57	0.55	0.54	0.53	0.50
100	0.79	0.54	0.50	0.49	0.48	0.46	0.44
200	0.60	0.39	0.37	0.36	0.35	0.33	0.32
500	0.46	0.28	0.26	0.26	0.26	0.25	0.23
1000	0.36	0.22	0.21	0.21	0.20	0.20	0.20

Table 31: Average excess over best known solutions over the larger AP instances for different number of offspring λ and iterations κ. In these experiments, the Don't Look Markers were used.

	CPU time [s]						
λ	$\kappa = 10$	$\kappa = 20$	$\kappa = 50$	$\kappa = 100$	$\kappa = 200$	$\kappa = 500$	$\kappa = 1000$
10	3	5	11	19	37	89	177
20	5	10	22	39	74	180	355
50	14	26	54	97	185	446	882
100	28	53	109	197	373	903	1786
200	55	106	217	393	746	1804	3567
500	139	267	545	988	1876	4538	8976
1000	276	525	1072	1950	3705	8974	17754

Table 32: Average CPU times over the larger AP instances for different number of offspring λ and iterations κ. In these experiments, the Don't Look Markers were used.

	Excess over best known solution [%]						
λ	$\kappa = 10$	$\kappa = 20$	$\kappa = 50$	$\kappa = 100$	$\kappa = 200$	$\kappa = 500$	$\kappa = 1000$
10	1.23	0.87	0.84	0.81	0.77	0.69	0.64
20	0.95	0.66	0.60	0.58	0.56	0.52	0.50
50	0.65	0.43	0.42	0.41	0.40	0.38	0.36
100	0.48	0.30	0.29	0.29	0.28	0.27	0.25
200	0.34	0.24	0.23	0.22	0.22	0.21	0.21
500	0.23	0.14	0.14	0.14	0.14	0.14	0.13
1000	0.16	0.10	0.10	0.10	0.10	0.10	0.10

Table 33: Average excess over best known solutions over the larger AP instances for different number of offspring λ and iterations κ. In these experiments, the Don't Look Markers were not used.

	CPU time [s]						
λ	$\kappa = 10$	$\kappa = 20$	$\kappa = 50$	$\kappa = 100$	$\kappa = 200$	$\kappa = 500$	$\kappa = 1000$
10	7	13	23	40	73	171	335
20	14	26	47	80	146	345	676
50	35	64	117	200	364	858	1 679
100	70	127	233	398	727	1 713	3 356
200	142	257	470	808	1 482	3 503	6 875
500	350	628	1 152	1 978	3 633	8 598	16 870
1000	699	1 252	2 293	3 947	7 255	17 177	33 717

Table 34: Average CPU times over the larger AP instances for different number of offspring λ and iterations κ. In these experiments, the Don't Look Markers were not used.

A.3   MINIMUM ENERGY BROADCAST

A.3.1   *Modified MIP Formulation*

For finding optimal solutions or lower bounds for the Minimum Energy Broadcast (MEB) instances, we have resorted to a modified Mixed Integer Programming (MIP) formulation based on the formulation presented by Montemanni *et al.* in [124]. The model is based on a flow formulation in a tree. The source node sends $n - 1$ units of flow, which are transported along the edges of the tree. Each other node has a demand of 1 unit of flow.

sorting   In a pre-processing step, Montemanni *et al.* sort all nodes according to their distances. In a way, they create a vector a'_i containing all nodes in the network ordered by non-decreasing distance to node i. This is done for all nodes $i \in V$. However, they do not use this vector a'_i directly in their formulation, but define an ancestor function a^i_j that gives the node that is one step closer to node i than node j. If no such node exists, node i itself is returned. If two nodes have the same distance to a node i, an arbitrary choice is made to create a fixed order.

Using their formulation, the MEB is the problem of minimizing the cost function C in this MIP formulation:

$$\text{min: } C = \sum_{(i,j) \in E} d_{ij}^{\alpha} \cdot y_{ij} \tag{A.8}$$

s.t.

$$y_{ij} \leq y_{ia^i_j} \qquad\qquad \forall (i,j) \in E, a^i_j \neq i \tag{A.9}$$

$$x_{ij} \leq (n-1) \cdot y_{ij} \qquad\qquad \forall (i,j) \in E \tag{A.10}$$

$$x_{ij} \in \mathbb{R} \qquad\qquad \forall (i,j) \in E \tag{A.11}$$

$$y_{ij} \in \{0,1\} \qquad\qquad \forall (i,j) \in E \tag{A.12}$$

$$\sum_{(i,j) \in E} x_{ij} - \sum_{(k,i) \in E} x_{ki} = \begin{cases} n-1 & \text{if } i = s \\ -1 & \text{otherwise} \end{cases} \qquad \forall i \in V \tag{A.13}$$

The model uses binary variables y_{ij} to denote that node i has sufficient transmission power to reach node j. Constraint (A.9) ensures that all closer nodes can also be reached. It also uses variables x_{ij} to denote the amount of flow along this edge. These variables define the broadcast tree. Constraint (A.10) ensures that only those edges are used for the broadcast that can be established using the transmission power setting y_{ij}. Constraint (A.13) gives a flow conservation constraint. It also defines the amount of flow emanating from the source node as $n - 1$, and the amount of flow absorbed by each other node as 1. This flow constraint ensures that a topology is created that spans the whole network.

In their model, the flow on an established edge is bounded by constraint (A.10). A simple inequation $x_{ij} \leq y_{ij}$ does not suffice here, as the flow on an edge may be larger than one. Nevertheless, this bound *improving step* can be tightened slightly. Only the source node needs to be allowed to transmit $n - 1$ units of flow over one edge. Every other node does not need to forward all these messages, since the node itself takes one message. So, every node except the source node only needs to be allowed to transmit $n - 2$ units of flow over one edge. The new bound can be expressed by these constraints:

$$x_{ij} \leq (n - 1) \cdot y_{ij} \qquad \forall i, j \in N : i = \text{source}$$
$$x_{ij} \leq (n - 2) \cdot y_{ij} \qquad \forall i, j \in N : i \neq \text{source}$$

This bound is tight in the sense that there is an instance where this bound is met. Consider a simple MEB problem with $n = 3$ nodes in *it's tight* a straight line, such that no node is placed in the same position as any other node. The first node is the source node. Such an instance is shown in Figure 59. The optimal solution to this MEB problem for $\alpha = 2$ is to have the source node transmit to the middle node which relays the message to the third node. The flow x in this case is as follows: The source node sends $n - 1 = 2$ units of flow to the second node, and the second node sends $n - 2 = 1$ unit of flow to the third node.

Using this modified formulation, we were able to speed up the calculation time for proving the optimality of a solution. In one case, the *speed-up* speed-up was about 50 %. This allowed us to prove the optimality of more instances from the 50 nodes set than before. All optimal solutions of this set and the CPU time used by CPlex to prove their optimality are shown in Table 35.

Besides reducing the running time, the modified formulation brings a second advantage. Its Linear Programming (LP) relaxation gives bet- *better* ter lower bounds. This is used in the Nested Partitioning (NP) in Sec- *relaxations* tion 5.3.2.

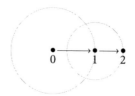

Figure 59: Tight example for the improved MIP formulation. Three wireless nodes are placed on a straight line. The solution for the MEB is shown.

Instance	Optimum	Time	Instance	Optimum	Time
p50.00	399 074.64	37 h	p50.15	371 694.65	47 h
p50.01	373 565.15	5 days	p50.16	414 587.42	22 h
p50.02	393 641.09	3 days	p50.17	355 937.07	3 days
p50.03	316 801.09	31 h	p50.18	376 617.33	5 days
p50.04	325 774.22	16 h	p50.19	335 059.72	82 min
p50.05	382 235.90	28 h	p50.20	414 768.96	15 days
p50.06	384 438.46	12 h	p50.21	361 354.27	12 h
p50.07	401 836.85	6 days	p50.22	329 043.51	17 min
p50.08	334 418.45	45 h	p50.23	383 321.04	2 days
p50.09	346 732.05	39 h	p50.24	404 855.92	3 days
p50.10	416 783.45	44 h	p50.25	363 200.32	105 s
p50.11	369 869.41	34 h	p50.26	406 631.51	25 h
p50.12	392 326.01	15 h	p50.27	451 059.62	16 days
p50.13	400 563.83	45 h	p50.28	415 832.44	3 days
p50.14	388 714.91	3 min	p50.29	380 492.77	3 days

Table 35: Optimal solutions for the MEB instances with $n = 50$ nodes. These solutions were proven to be optimal using the modified MIP formulation and CPlex 10.1.

Instance	Optimum	Instance	Optimum	Instance	Optimum
p20.00	407 250.81	p20.10	316 734.39	p20.20	403 582.74
p20.01	446 905.52	p20.11	289 200.92	p20.21	271 958.28
p20.02	335 102.42	p20.12	314 511.98	p20.22	328 659.78
p20.03	488 344.90	p20.13	346 234.51	p20.23	326 654.08
p20.04	516 117.75	p20.14	301 426.68	p20.24	395 859.67
p20.05	300 869.14	p20.15	457 467.93	p20.25	453 517.28
p20.06	250 553.15	p20.16	484 437.68	p20.26	461 547.18
p20.07	347 454.08	p20.17	380 175.41	p20.27	389 057.00
p20.08	390 795.34	p20.18	320 300.23	p20.28	279 251.95
p20.09	447 659.11	p20.19	461 267.52	p20.29	299 586.76

Table 36: Optimal solutions for the MEB instances with $n = 20$ nodes. These solutions were proven to be optimal using the original MIP formulation by Montemanni *et al.* and CPlex 10.1.

Table 36 gives the optimal values for all 20 nodes problems. The original MIP formulation was used for these instances, as CPlex took only a couple of seconds to prove their optimality.

A.3.2   Parameter Study for the NP and the ACO for the MPSCP

Tables 38, 37 and 39 show average results for the NP, the NP+MMAS, and the pure MMAS, using different values for the parameters q (probability to pick the best arc when generating a new sample), s (sample size), a (influence of pheromones), and b (influence of additional cost). These experiments were conducted on all problems of the 50 nodes set.

It can clearly be seen that the sample rate s has the largest influence on the quality of the solutions. However, for some parameter settings the quality actually degrades when increasing the sample rate past $s = 200$. This is the case for $a = 2$, $b = 2$ and $q \leq 0.8$.

Setting $a = 0$ means calling NP without pheromone information. Here, the best values are found with higher q values. This means, that *pure NP* selecting the arcs with lower additional costs is a good choice when no information from pheromones is available. However, using $b = 2$ leads to outstandingly bad solutions. Here, the probability of choos-

$a = 0\ b = 1$	$s = 10$	$s = 20$	$s = 50$	$s = 100$	$s = 200$	$s = 500$	$s = 1000$
$q = 0.1$	6.34 %	5.76 %	4.91 %	4.35 %	3.68 %	3.07 %	2.69 %
$q = 0.2$	6.21 %	5.73 %	4.75 %	4.10 %	3.63 %	2.96 %	2.53 %
$q = 0.5$	5.50 %	4.95 %	4.12 %	3.53 %	3.00 %	2.34 %	1.89 %
$q = 0.8$	3.86 %	3.30 %	2.58 %	2.12 %	1.68 %	1.19 %	0.93 %
$q = 0.9$	3.17 %	2.80 %	2.19 %	1.81 %	1.46 %	1.06 %	0.82 %

$a = 0\ b = 2$	$s = 10$	$s = 20$	$s = 50$	$s = 100$	$s = 200$	$s = 500$	$s = 1000$
$q = 0.1$	8.12 %	7.78 %	6.87 %	6.13 %	5.87 %	5.14 %	4.62 %
$q = 0.2$	8.29 %	7.84 %	6.90 %	6.37 %	5.87 %	5.23 %	4.73 %
$q = 0.5$	8.76 %	8.42 %	7.53 %	6.86 %	6.34 %	5.74 %	5.24 %
$q = 0.8$	10.11 %	9.66 %	8.71 %	8.26 %	7.50 %	6.81 %	6.33 %
$q = 0.9$	10.83 %	10.42 %	9.65 %	9.12 %	8.45 %	7.55 %	7.02 %

Table 37: Average results from the parameter analysis for the NP without pheromones. Sample rate s, influence of pheromones a, influence of additional cost b, and probability to select the best arc q. The values are the average excess over the best known solutions for all instances of the 50 nodes set.

ing longer arcs is decreased to a degree where good solutions are impossible to reach.

NP+MMAS

When using pheromone information, the best solutions are found with $a = 1$, $b = 2$ and $q \in \{0.1, 0.2, 0.5\}$ and $s \in \{500, 1000\}$. In general, $q = 0.5$ is a good choice. This gives a good balance between choosing low cost arcs and still allowing the ant system to explore other solutions.

$a = 1\ b = 1$	$s = 10$	$s = 20$	$s = 50$	$s = 100$	$s = 200$	$s = 500$	$s = 1000$
$q = 0.1$	6.26 %	5.81 %	4.68 %	3.75 %	2.31 %	1.25 %	1.02 %
$q = 0.2$	6.11 %	5.51 %	4.49 %	3.49 %	2.18 %	1.29 %	1.06 %
$q = 0.5$	5.35 %	4.89 %	3.68 %	2.61 %	1.49 %	1.24 %	0.96 %
$q = 0.8$	3.72 %	3.23 %	2.00 %	1.13 %	0.74 %	0.80 %	0.75 %
$q = 0.9$	3.15 %	2.55 %	1.64 %	0.95 %	0.66 %	0.72 %	0.72 %

$a = 1\ b = 2$	$s = 10$	$s = 20$	$s = 50$	$s = 100$	$s = 200$	$s = 500$	$s = 1000$
$q = 0.1$	3.33 %	2.71 %	1.72 %	1.00 %	0.31 %	0.21 %	0.17 %
$q = 0.2$	3.13 %	2.55 %	1.53 %	0.81 %	0.31 %	0.21 %	0.18 %
$q = 0.5$	2.63 %	2.20 %	1.16 %	0.57 %	0.30 %	0.24 %	0.17 %
$q = 0.8$	2.54 %	2.02 %	1.26 %	0.71 %	0.41 %	0.30 %	0.30 %
$q = 0.9$	2.98 %	2.51 %	1.69 %	1.13 %	0.67 %	0.55 %	0.41 %

$a = 2\ b = 2$	$s = 10$	$s = 20$	$s = 50$	$s = 100$	$s = 200$	$s = 500$	$s = 1000$
$q = 0.1$	3.32 %	2.56 %	1.29 %	0.50 %	0.34 %	0.42 %	0.36 %
$q = 0.2$	3.08 %	2.43 %	1.18 %	0.47 %	0.39 %	0.41 %	0.38 %
$q = 0.5$	2.47 %	1.98 %	0.87 %	0.42 %	0.41 %	0.42 %	0.38 %
$q = 0.8$	2.46 %	1.95 %	0.99 %	0.56 %	0.56 %	0.69 %	0.60 %
$q = 0.9$	2.85 %	2.45 %	1.47 %	0.89 %	0.91 %	0.89 %	0.95 %

Table 38: Average results from the parameter analysis for the NP with ACO. Sample rate s, influence of pheromones a, influence of additional cost b, and probability to select the best arc q. The values are the average excess over the best known solutions for all instances of the 50 nodes set.

$a = 1\ b = 1$	$s = 10$	$s = 20$	$s = 50$	$s = 100$	$s = 200$	$s = 500$	$s = 1000$
$q = 0.1$	6.77 %	5.71 %	4.50 %	3.37 %	1.45 %	0.79 %	0.60 %
$q = 0.2$	6.65 %	5.51 %	4.35 %	3.18 %	1.25 %	0.69 %	0.54 %
$q = 0.5$	5.57 %	4.61 %	3.32 %	1.99 %	0.73 %	0.52 %	0.47 %
$q = 0.8$	3.34 %	2.57 %	1.49 %	0.57 %	0.31 %	0.26 %	0.20 %
$q = 0.9$	2.74 %	2.07 %	1.19 %	0.53 %	0.31 %	0.29 %	0.24 %

$a = 1\ b = 2$	$s = 10$	$s = 20$	$s = 50$	$s = 100$	$s = 200$	$s = 500$	$s = 1000$
$q = 0.1$	2.76 %	2.01 %	1.18 %	0.48 %	0.14 %	0.09 %	0.04 %
$q = 0.2$	2.51 %	1.86 %	1.04 %	0.43 %	0.14 %	0.08 %	0.07 %
$q = 0.5$	2.07 %	1.45 %	0.72 %	0.28 %	0.14 %	0.10 %	0.09 %
$q = 0.8$	2.06 %	1.57 %	0.86 %	0.44 %	0.20 %	0.16 %	0.14 %
$q = 0.9$	2.49 %	2.01 %	1.35 %	0.80 %	0.35 %	0.26 %	0.21 %

$a = 2\ b = 2$	$s = 10$	$s = 20$	$s = 50$	$s = 100$	$s = 200$	$s = 500$	$s = 1000$
$q = 0.1$	2.68 %	1.80 %	0.78 %	0.19 %	0.14 %	0.14 %	0.15 %
$q = 0.2$	2.46 %	1.74 %	0.66 %	0.20 %	0.17 %	0.17 %	0.15 %
$q = 0.5$	2.04 %	1.37 %	0.47 %	0.20 %	0.15 %	0.15 %	0.18 %
$q = 0.8$	1.97 %	1.43 %	0.62 %	0.28 %	0.25 %	0.25 %	0.22 %
$q = 0.9$	2.54 %	1.88 %	1.07 %	0.50 %	0.42 %	0.39 %	0.37 %

Table 39: Average results from the parameter analysis for the ACO without NP. Sample rate s, influence of pheromones a, influence of additional cost b, and probability to select the best arc q. The values are the average excess over the best known solutions for all instances of the 50 nodes set.

B

BIBLIOGRAPHY

[1] Ravindra K. Ahuja and V. V. S. Murty. Exact and Heuristic Algorithms for the Optimum Communication Spanning Tree Problem. *Transportation Science*, 21(3):163–170, 1987. (Cited on pages 39 and 42.)

[2] Ernst Althaus, Gruia Călinescu, Ion I. Măndoiu, Sushil K. Prasad, Nickolay Tchervenski, and Alexander Zelikovsky. Power Efficient Range Assignment in Ad-Hoc Wireless Networks. In *Proc. of the IEEE Wireless Communications and Networking Conference (WCNC 2003)*, pages 1889–1894. IEEE Computer Society Press, Los Alamitos, 2003. (Cited on pages 188, 190, 192, 199, and 206.)

[3] Ernst Althaus, Gruia Călinescu, Ion I. Măndoiu, Sushil K. Prasad, Nickolay Tchervenski, and Alexander Zelikovsky. Power Efficient Range Assignment for Symmetric Connectivity in Static Ad Hoc Wireless Networks. *Wireless Networks*, 12(3):287–299, 2006. (Cited on pages 188, 190, 192, 199, and 206.)

[4] Christoph Ambühl. An Optimal Bound for the MST Algorithm to Compute Energy Efficient Broadcast Trees in Wireless Networks. In Luís Caires, Giuseppe F. Italiano, Luís Monteiro, Catuscia Palamidessi, and Moti Yung, editors, *ICALP 2005*, volume 3580 of *LNCS*, pages 1139–1150. Springer, Heidelberg, 2005. (Cited on pages 162 and 173.)

[5] David P. Anderson. BOINC: A System for Public-Resource Computing and Storage. In *Proceedings of the 5th IEEE/ACM International Workshop on Grid Computing*, pages 4–10, 2004. (Cited on page 137.)

[6] Nazareno Andrade, Francisco Vilar Brasileiro, Walfredo Cirne, and Miranda Mowbray. Automatic grid assembly by promoting collaboration in Peer-to-Peer grids. *Journal of Parallel and Distributed Computing*, 67(8):957–966, 2007. (Cited on page 137.)

[7] Sameh El-Ansary, Luc Onana Alima, Per Brand, and Seif Haridi. Efficient Broadcast in Structured P2P Networks. In M. Frans

Kaashoek and Ion Stoica, editors, *IPTPS 2003*, volume 2735 of *Lecture Notes in Computer Science*, pages 304–314. Springer, Heidelberg, 2003. (Cited on page 142.)

[8] W. Ross Ashby. Principles of the Self-Organizing Dynamic System. *The Journal of General Psychology*, 37:125–128, 1947. (Cited on page 30.)

[9] Thomas Bäck. *Evolutionary Algorithms in Theory and Practise.* Oxford University Press, New York, 1996. ISBN 978-0195099713. (Cited on page 16.)

[10] Thomas Bäck, Ulrich Hammel, and Hans-Paul Schwefel. Evolutionary Computation: Comments on the History and Current State. *IEEE Transactions on Evolutionary Computation*, 1(1):3–17, 1997. (Cited on pages 17 and 18.)

[11] Suman Banerjee, Timothy G. Griffin, and Marcelo Pias. The Interdomain Connectivity of PlanetLab Nodes. In Chadi Barakat and Ian Pratt, editors, *Proceedings of the 5th International Workshop on Passive and Active Network Measurement, (PAM 2004)*, volume 3015 of *Lecture Notes in Computer Science*, pages 73–82. Springer, 2004. (Cited on page 49.)

[12] Ingmar Baumgart, Bernhard Heep, and Stephan Krause. OverSim: A Flexible Overlay Network Simulation Framework. In *Proceedings of the Tenth IEEE Global Internet Symposium (GI'07) in conjunction with IEEE INFOCOM 2007*, 2007. (Cited on page 124.)

[13] Richard Bellman. On a Routing Problem. *Quarterly of Applied Mathematics*, 16(1):87–90, 1958. (Cited on pages 35 and 65.)

[14] Jon Louis Bentley. Experiments on traveling salesman heuristics. In *SODA'90: Proceedings of the first annual ACM-SIAM symposium on Discrete algorithms*, pages 91–99, Philadelphia, USA, 1990. Society for Industrial and Applied Mathematics. (Cited on pages 24 and 98.)

[15] Andrew Brampton, Andrew MacQuire, Idris A. Rai, Nicholas J. P. Race, and Laurent Mathy. Stealth Distributed Hash Table: A Robust and Flexible Super-Peered DHT. In Christophe Diot and

Mostafa H. Ammar, editors, *Proceedings of the 2006 ACM Conference on Emerging Network Experiment and Technology, CoNEXT 2006*. ACM, 2006. (Cited on page 49.)

[16] Eli Brosh and Yuval Shavitt. Approximation and Heuristic Algorithms for Minimum-Delay Application-Layer Multicast Trees. In *The 23rd International Conference on Computer Communications, IEEE INFOCOM*, Hong Kong, 2004. (Cited on page 62.)

[17] Mario Čagalj, Jean-Pierre Hubaux, and Christian Enz. Minimum-Energy Broadcast in All-Wireless Networks: NP-Completeness and Distribution Issues. In *MobiCom 2002: Proceedings of the 8th Annual International Conference on Mobile Computing and Networking*, pages 172–182, New York, 2002. ACM Press. ISBN 1-58113-486-X. (Cited on pages 159 and 163.)

[18] Gruia Călinescu, Ion I. Măndoiu, and Alexander Zelikovsky. Symmetric Connectivity with Minimum Power Consumption in Radio Networks. In Ricardo A. Baeza-Yates, Ugo Montanari, and Nicola Santoro, editors, *Proc. 2nd IFIP International Conference on Theoretical Computer Science*, volume 223 of *IFIP Conference Proceedings*, pages 119–130. Kluwer, Dordrecht, 2002. (Cited on pages 160, 161, 188, 189, 190, and 192.)

[19] Scott Camazine, Jean-Louis Deneubourg, Nigel R. Franks, James Sneyd, Guy Theraulaz, and Eric Bonabeau. *Self-Organization in Biological Systems*. Princeton Studies in Complexity. Princeton University Press, 2001. ISBN 978-0-691-01211-7. (Cited on page 30.)

[20] James F. Campbell. Integer programming formulations of discrete hub location problems. *European Journal of Operational Research*, 72:387–405, 1994. (Cited on page 89.)

[21] Rui Campos and Manuel Ricardo. A fast algorithm for computing minimum routing cost spanning trees. *Computer Networks*, 52(17):3229–3247, 2008. (Cited on pages 38 and 39.)

[22] Vladimír Černý. Thermodynamical Approach to the Travelling Salesman Problem: An Efficient Simulation Algorithm. *Journal of Optimization Theory and Applications*, 45(1):41–51, 1985. (Cited on page 24.)

[23] Arjav J. Chakravarti, Gerald Baumgartner, and Mario Lauria. The Organic Grid: Self-Organizing Computation on a Peer-to-Peer Network. *IEEE Transactions on Systems, Man, and Cybernetics, Part A*, 35(3):373–384, 2005. (Cited on page 137.)

[24] Ernest J. H. Chang. Echo Algorithms: Depth Parallel Operations on General Graphs. *IEEE Transactions on Software Engineering*, SE-8(4):391–401, 1982. (Cited on page 33.)

[25] Xiuzhen Cheng, Bhagirath Narahari, Rahul Simha, Maggie Xiaoyan Cheng, and Dan Liu. Strong Minimum Energy Topology in Wireless Sensor Networks: NP-Completeness and Heuristics. *IEEE Transactions on Mobile Computing*, 2(3):248–256, 2003. (Cited on pages 160, 161, 188, and 205.)

[26] Brent Chun, David Culler, Timothy Roscoe, Andy Bavier, Larry Peterson, Mike Wawrzoniak, and Mic Bowman. PlanetLab: An Overlay Testbed for Broad-Coverage Services. *ACM SIGCOMM Computer Communication Review*, 33(3):3–12, 2003. (Cited on page 48.)

[27] Andrea E. F. Clementi, Paolo Penna, and Riccardo Silvestri. Hardness Results for the Power Range Assignment Problem in Packet Radio Networks. In Dorit S. Hochbaum, Klaus Jansen, José D. P. Rolim, and Alistair Sinclair, editors, *RANDOM 1999 and APPROX 1999*, volume 1671 of *LNCS*, pages 197–208. Springer, Heidelberg, 1999. (Cited on page 188.)

[28] Andrea E. F. Clementi, Pilu Crescenzi, Paolo Penna, Gianluca Rossi, and Paola Vocca. On the Complexity of Computing Minimum Energy Consumption Broadcast Subgraphs. In Afonso Ferreira and Horst Reichel, editors, *STACS 2001*, volume 2010 of *LNCS*, pages 121–131. Springer, Heidelberg, 2001. (Cited on pages 159 and 162.)

[29] Stephen A. Cook. The Complexity of Theorem-Proving Procedures. In *Proceedings of the third annual ACM symposium on Theory of computing*, pages 151–158. ACM, New York, 1971. (Cited on page 13.)

[30] Manuel Costa, Miguel Castro, Antony Rowstron, and Peter Key. PIC: Practical Internet Coordinates for Distance Estimation. In

Proceedings of the 24th IEEE International Conference on Distributed Computing Systems (ICDCS'04), pages 178–187. IEEE Press, 2004. (Cited on page 120.)

[31] George F. Coulouris and Jean Dollimore. *Distributed Systems: Concepts and Design*. International Computer Science Series. Addison Wesley, 1988. ISBN 0-201-18059-6. (Cited on page 28.)

[32] Russ Cox, Frank Dabek, Franz Kaashoek, Jinyang Li, and Robert Morris. Practical, Distributed Network Coordinates. In *Proceedings of the ACM HotNets Workshop*, 2003. (Cited on pages 119 and 120.)

[33] Russ Cox, Frank Dabek, M. Frans Kaashoek, Jinyang Li, and Robert Morris. Practical, Distributed Network Coordinates. *ACM SIGCOMM Computer Communication Review*, 34(1):113–118, 2004. (Cited on pages 119, 120, and 132.)

[34] Nichael Lynn Cramer. A Representation for the Adaptive Generation of Simple Sequential Programs. In John J. Grefenstette, editor, *Proceedings of the 1st International Conference on Genetic Algorithms and Their Applications*, pages 183–187. L. Erlbaum Associates Inc., Hillsdale, 1985. (Cited on page 16.)

[35] Frank Dabek, Russ Cox, M. Frans Kaashoek, and Robert Morris. Vivaldi: A Decentralized Network Coordinate System. In *ACM SIGCOMM'04*, pages 15–26, 2004. (Cited on page 120.)

[36] Charles Darwin. *The Origin of Species*. Classic Reprint. Forgotten Books, 2007. ISBN 978-1-4400-3807-5. First published in 1872. (Cited on page 17.)

[37] Arindam K. Das, Robert J. Marks, Mohamed El-Sharkawi, Payman Arabshahi, and Andrew Gray. The Minimum Power Broadcast Problem in Wireless Networks: An Ant Colony System Approach. In *IEEE CAS Workshop on Wireless Communications and Networking*, 2002. (Cited on pages 163 and 164.)

[38] Arindam K. Das, Robert J. Marks, Mohamed El-Sharkawi, Payman Arabshahi, and Andrew Gray. Minimum Power Broadcast Trees for Wireless Networks: Integer Programming Formulations. In *Proceedings of the 22nd IEEE INFOCOM 2003*, pages 1001–1010, 2003. (Cited on page 164.)

[39] Arindam K. Das, Robert J. Marks, Mohamed El-Sharkawi, Payman Arabshahi, and Andrew Gray. r-shrink: A Heuristic for Improving Minimum Power Broadcast Trees in Wireless Networks. In *Global Telecommunications Conference, GLOBECOM 2003*, pages 523–527, Los Alamitos, 2003. IEEE. (Cited on pages 162, 163, 168, 174, and 186.)

[40] Kenneth A. De Jong. Are Genetic Algorithms Function Optimizers? In Reinhard Männer and Bernard Manderick, editors, *Parallel Problem Solving from Nature 2, PPSN-II*, pages 3–13. Elsevier, 1992. (Cited on page 16.)

[41] Hermann de Meer and Christian Koppen. Characterization of Self-Organization. In Ralf Steinmetz and Klaus Wehrle, editors, *Peer-to-Peer Systems and Applications*, volume 3485 of *Lecture Notes in Computer Science*, pages 227–246. Springer, 2005. (Cited on page 31.)

[42] Alan J. Demers, Daniel H. Greene, Carl Hauser, Wes Irish, John Larson, Scott Shenker, Howard E. Sturgis, Daniel C. Swinehart, and Douglas B. Terry. Epidemic Algorithms for Replicated Database Maintenance. In *Proceedings of the 6th Annual ACM Symposium on Principles of Distributed Computing (PODC)*, pages 1–12, 1987. (Cited on page 142.)

[43] Nihan Çetin Demirel and M. Duran Toksarı. Optimization of the quadratic assignment problem using an ant colony algorithm. *Applied Mathematics and Computation*, 183:427–435, 2006. (Cited on page 26.)

[44] Edsger W. Dijkstra. A Note on Two Problems in Connexion with Graphs. *Numerische Mathematik*, 1:269–271, 1959. (Cited on pages 6, 35, and 65.)

[45] Enrique Domínguez, José Muñoz, and Enrique Mérida. A recurrent neural network model for the p-hub problem. In José Mira and José R. Álvarez, editors, *IWANN 2003: International Work-Conference on Artificial and Natural Neural Networks*, volume 2687 of *LNCS*, pages 734–741. Springer, 2003. (Cited on page 92.)

[46] Benoit Donnet, Bamba Gueye, and Mohamed Ali Kaafar. A Survey on Network Coordinates Systems, Design, and Secur-

ity. *IEEE Communication Surveys & Tutorials*, 2010. To appear. (Cited on page 117.)

[47] Marco Dorigo and Thomas Stützle. *Ant Colony Optimization*. MIT Press, Cambridge, 2004. ISBN 0-262-04219-3. (Cited on pages 26, 169, and 170.)

[48] Marco Dorigo, Vittorio Maniezzo, and Alberto Colorni. Ant System: An Autocatalytic Optimizing Process. Technical Report 91-016, Politecnico di Milano, 1991. (Cited on page 25.)

[49] Marco Dorigo, Vittorio Maniezzo, and Alberto Colorni. Ant System: Optimization by a Colony of Cooperating Agents. *IEEE Transactions on Systems, Man, and Cybernetics, Part B: Cybernetics*, 26(1):29–41, 1996. (Cited on page 25.)

[50] Jamie Ebery. Solving large single allocation p-hub problems with two or three hubs. *European Journal of Operational Research*, 128 (2):447–458, 2001. (Cited on page 91.)

[51] Andreas T. Ernst and Mohan Krishnamoorthy. Efficient Algorithms for the Uncapacitated Single Allocation p-Hub Median Problem. *Location Science*, 4(3):139–154, 1996. (Cited on pages 89, 91, 99, and 221.)

[52] Andreas T. Ernst and Mohan Krishnamoorthy. An Exact Solution Approach Based on Shortest-Paths for p-Hub Median Problems. *INFORMS Journal on Computing*, 10(2):149–162, 1998. (Cited on page 89.)

[53] Patrick Thomas Eugster, Rachid Guerraoui, Anne-Marie Kermarrec, and Laurent Massoulie. From Epidemics to Distributed Computing. *IEEE Computer*, 37(5):60–67, 2004. (Cited on page 66.)

[54] Matteo Fischetti, Giuseppe Lancia, and Paolo Serafini. Exact Algorithms for Minimum Routing Cost Trees. *Networks*, 39(3):161–173, 2002. (Cited on pages 39 and 60.)

[55] Lester R. Ford, Jr. and Delbert R. Fulkerson. *Flows in Networks*. Princeton University Press, 1962. (Cited on pages 35 and 65.)

[56] Pedro García, Carles Pairot, Rubén Mondéjar, Jordi Pujol, Helio Tejedor, and Robert Rallo. PlanetSim: A New Overlay Network Simulation Framework. In Thomas Gschwind and Cecilia Mascolo, editors, *Software Engineering and Middleware*, volume 3437 of *Lecture Notes in Computer Science*, pages 123–136. Springer, Heidelberg, 2005. (Cited on page 124.)

[57] Michael R. Garey and David S. Johnson. *Computers and Intractibility: A guide to the theory of NP-completeness*. W. H. Freeman and Co, San Francisco, 1979. ISBN 0-7167-1044-7. (Cited on pages 10, 12, 13, 29, 34, 39, 188, and 223.)

[58] Michel Gendreau, Alain Hertz, and Gilbert Laporte. New Insertion and Postoptimization Procedures for the Traveling Salesman Problem. *Operations Research*, 40(6):1086–1094, 1992. (Cited on page 21.)

[59] Fred Glover. Future paths for integer programming and links to artificial intelligence. *Computers and Operations Research*, 13(5): 533–549, 1986. (Cited on page 23.)

[60] Krishna P. Gummadi, Stefan Saroiu, and Steven D. Gribble. King: Estimating Latency between Arbitrary Internet End Hosts. In *IMW'02: Proceedings of the 2nd ACM SIGCOMM Workshop on Internet Measurement*, pages 5–18, New York, USA, 2002. ACM. (Cited on page 125.)

[61] P. Krishna Gummadi, Ramakrishna Gummadi, Steven D. Gribble, Sylvia Ratnasamy, Scott Shenker, and Ian Stoica. The impact of DHT routing geometry on resilience and proximity. In *SIGCOMM'03: Proceedings of the 2003 conference on Applications, technologies, architectures, and protocols for computer communications*, pages 381–394, New York, USA, 2003. ACM. (Cited on pages 138 and 144.)

[62] Song Guo and Oliver Yang. Localized Operations for Distributed Minimum Energy Multicast Algorithm in Mobile Ad Hoc Networks. *IEEE Transactions on Parallel and Distributed Systems*, 18(2), 2007. (Cited on page 206.)

[63] Song Guo and Oliver W. W. Yang. Energy-aware multicasting in wireless ad hoc networks: A survey and discussion. *Computer*

Communications, 30(9):2129–2148, 2007. (Cited on pages 164 and 172.)

[64] Yuri Gurevich, Philipp W. Kutter, Martin Odersky, and Lothar Thiele, editors. *Abstract State Machines, Theory and Applications, International Workshop, ASM 2000, Monte Verità, Switzerland, March 19-24, 2000, Proceedings*, volume 1912 of *Lecture Notes in Computer Science*, 2000. Springer. ISBN 3-540-67959-6. (Cited on page 30.)

[65] Pierre Hansen and Nenad Mladenović. First vs. best improvement: An empirical study. *Discrete Applied Mathematics*, 154(5): 802–817, 2006. (Cited on page 20.)

[66] Sabine Hauert, Jean-Christophe Zufferey, and Dario Floreano. Evolved swarming without positioning information: an application in aerial communication relay. *Autonomous Robots*, 26(1): 21–32, 2009. (Cited on page 156.)

[67] Hugo Hernández, Christian Blum, and Guillem Francès. Ant Colony Optimization for Energy-Efficient Broadcasting in Ad-Hoc Networks. In Marco Dorigo, Mauro Birattari, Christian Blum, Maurice Clerc, Thomas Stützle, and Alan F. T. Winfield, editors, *ANTS 2008*, volume 5217 of *LNCS*, pages 25–36. Springer, Heidelberg, 2008. (Cited on pages 164 and 187.)

[68] Francis Heylighen. The Science of Self-Organization and Adaptivity. In *Knowledge Management, Organizational Intelligence and Learning, and Complexity*, The Encyclopedia of Life Support Systems, EOLSS, pages 253–280. Publishers Co. Ltd, 1999. (Cited on page 31.)

[69] John H. Holland. Outline for a Logical Theory of Adaptive Systems. *Journal of the ACM*, 9(3):297–314, 1962. (Cited on page 18.)

[70] Feng Hong, Minglu Li, Xinda Lu, Jiadi Yu, Yi Wang, and Ying Li. HP-Chord: A Peer-to-Peer Overlay to Achieve Better Routing Efficiency by Exploiting Heterogeneity and Proximity. In *Proceedings of the 3rd International Conference on Grid and Cooperative Computing*, volume 3251 of *Lecture Notes in Computer Science*, pages 626–633. Springer, 2004. (Cited on page 138.)

[71] Holger H. Hoos and Thomas Stützle. *Stochastic Local Search: Foundations and Applications.* The Morgan Kaufmann Series in Artificial Intelligence. Morgan Kaufmann, 2004. ISBN 1-55860-872-9. (Cited on pages 22 and 172.)

[72] Juraj Hromkovič. *Algorithmics for Hard Problems: Introduction to Combinatorial Optimization, Randomization, Approximation, and Heuristics.* Springer, Berlin, 2004. ISBN 3-540-44134-4. 2nd edition. (Cited on page 15.)

[73] T. C. Hu. Optimum Communication Spanning Trees. *SIAM Journal of Computing,* 3(3):188–195, 1974. (Cited on pages 34 and 38.)

[74] Adriana Iamnitchi and Ian T. Foster. A Peer-to-Peer Approach to Resource Location in Grid Environments. In Jan Weglarz, Jarek Nabrzyski, Jennifer M. Schopf, and M. Stroinski, editors, *Grid Resource Management: State of the Art and Future Trends,* pages 413–429. Kluwer Publishing, Dordrecht, 2004. (Cited on page 137.)

[75] Aleksandar Ilić, Dragan Urošević, Jack Brimberg, and Nenad Mladenović. A general variable neighborhood search for solving the uncapacitated single allocation p-hub median problem. *European Journal of Operational Research,* 206(2):289–300, 2010. To appear. (Cited on pages 92, 94, 104, 106, and 107.)

[76] ILOG S.A. *ILOG CPLEX 10.1 User's Manual.* Gentilly Cedex, France, and Mountain View, USA, 2006. URL http://www.cplex.com/. (Cited on pages 100, 104, 127, 164, 176, 177, 180, and 199.)

[77] Information Sciences Institute, University of Southern California. The Network Simulator – ns-2. Website, 2009. URL http://www.isi.edu/nsnam/ns/. (Cited on page 124.)

[78] Samuel L. S. Jacoby, Janusz S. Kowalik, and Joseph T. Pizzo. *Iterative Methods for Nonlinear Optimization Problems.* Prentice-Hall, Englewood Cliffs, 1972. ISBN 0-13-508119-X. (Cited on page 120.)

[79] Márk Jelasity, Wojtek Kowalczyk, and Maarten van Steen. Newscast Computing. Technical Report IR-CS-006, Vrije Universiteit Amsterdam, Department of Computer Science, Amsterdam, The Netherlands, 2003. (Cited on page 111.)

[80] Gian Paolo Jesi, Alberto Montresor, and Özalp Babaoglu. Proximity-Aware Superpeer Overlay Topologies. In Alexander Keller and Jean-Philippe Martin-Flatin, editors, *Proceedings of SelfMan'06*, volume 3996 of *Lecture Notes in Computer Science*, pages 43–57. Springer, 2006. (Cited on pages 85, 111, and 138.)

[81] David S. Johnson, Jan K. Lenstra, and Alexander H. G. Rinnooy Kan. The Complexity of the Network Design Problem. *Networks*, 8:279–285, 1978. (Cited on pages 34, 38, and 39.)

[82] Bryant A. Julstrom. The blob code is competitive with edge-sets in genetic algorithms for the minimum routing cost spanning tree problem. In *GECCO'05: Proceedings of the 2005 conference on Genetic and evolutionary computation*, pages 585–590, Washington DC, USA, 2005. ACM Press. ISBN 1-59593-010-8. (Cited on page 39.)

[83] Intae Kang and Radha Poovendran. Broadcast with Heterogeneous Node Capability. In *Global Telecommunications Conference, GLOBECOM 2004*, pages 4114–4119, Los Alamitos, 2004. IEEE. (Cited on page 163.)

[84] Intae Kang and Radha Poovendran. Iterated Local Optimization for Minimum Energy Broadcast. In *3rd International Symposium on Modeling and Optimization in Mobile, Ad-Hoc and Wireless Networks (WiOpt)*, pages 332–341, Los Alamitos, 2005. IEEE Computer Society. ISBN 0-7695-2267-X. (Cited on pages 163, 178, and 179.)

[85] Jamal N. Al-Karaki, Raza Ul-Mustafa, and Ahmed E. Kamal. Data aggregation and routing in Wireless Sensor Networks: Optimal and heuristic algorithms. *Computer Networks: The International Journal of Computer and Telecommunications Networking*, 53(7):945–960, 2009. (Cited on page 159.)

[86] Holger Karl and Andreas Willig. *Protocols and Architectures for Wireless Sensor Networks*. John Wiley & Sons, Chichester, 2005. ISBN 978-0470095102. (Cited on pages 7, 156, 160, and 161.)

[87] S. Kirkpatrick, C. D. Gelatt, Jr., and M. P. Vecchi. Optimization by Simulated Annealing. *Science*, 220(4598):671–680, 1983. (Cited on page 24.)

[88] Lefteris M. Kirousis, Evangelos Kranakis, Danny Krizanc, and Andrzej Pelc. Power Consumption in Packet Radio Networks. *Theoretical Computer Science*, 243(1–2):289–305, 2000. (Cited on page 188.)

[89] Michael Kleis, Eng Keong Lua, and Xiaoming Zhou. Hierarchical Peer-to-Peer Networks Using Lightweight SuperPeer Topologies. In *Proceedings of the 10th IEEE Symposium on Computers and Communications (ISCC 2005)*, pages 143–148, 2005. (Cited on page 111.)

[90] Derrick Kondo, Michela Taufer, Charles L. Brooks, Henri Casanova, and Andrew A. Chien. Characterizing and Evaluating Desktop Grids: An Empirical Study. In *Proceedings of the 18th International Parallel and Distributed Processing Symposium, IPDPS 2004*, 2004. (Cited on page 137.)

[91] John R. Koza. Genetic Programming: A Paradigm for Genetically Breeding Populations of Computer Programs to Solve Problems. Technical Report STAN-CS-90-1314, Stanford University Computer Science Department, 1990. (Cited on pages 16 and 18.)

[92] Jozef Kratica, Zorica Stanimirović, Dušan Tošić, and Vladimir Filipović. Two genetic algorithms for solving the uncapacitated single allocation p-hub median problem. *European Journal of Operational Research*, 182(1):15–28, 2007. (Cited on pages 92 and 107.)

[93] Jozef Kratica, Jelena Kojić, Dušan Tošić, Vladimir Filipović, and Djordje Dugošija. Two hybrid genetic algorithms for solving the super-peer selection problem. In *WSC2008 conference, Advances in Soft Computing*. Springer, 2009. (Cited on page 92.)

[94] Joseph B. Kruskal, Jr. On the Shortest Spanning Subtree of a Graph and the Traveling Salesman Problem. *Proceedings of the American Mathematical Society*, 7(1):48–50, 1956. (Cited on pages 6, 13, 37, and 189.)

[95] Gerald Kunzmann, Robert Nagel, Tobias Hoßfeld, Andreas Binzenhöfer, and Kolja Eger. Efficient Simulation of Large-Scale P2P Networks: Modeling Network Transmission Times. In *Workshop on Modeling, Simulation and Optimization of Peer-to-peer*

environments (MSOP2P) in conjunction with Euromicro (PDP 2007), pages 475–481, Naples, Italy, 2007. (Cited on page 124.)

[96] Eugene L. Lawler and D. E. Wood. Branch-and-Bound Methods: A Survey. *Operations Research*, 14(4):699–719, 1966. (Cited on page 14.)

[97] Katharina A. Lehmann and Michael Kaufmann. Evolutionary Algorithms for the Self-Organized Evolution of Networks. In *Proceedings of the 2005 conference on Genetic and evolutionary computation*, pages 563–570, New York, USA, 2005. ACM. (Cited on page 62.)

[98] Dongsheng Li, Nong Xiao, and Xicheng Lu. Topology and Resource Discovery in Peer-to-Peer Overlay Networks. In *Grid and Cooperative Computing – GCC 2004 Workshops*, volume 3252 of *Lecture Notes in Computer Science*, pages 221–228. Springer, 2004. (Cited on page 85.)

[99] Jinyang Li, Jeremy Stribling, Robert Morris, M. Frans Kaashoek, and Thomer M. Gil. A performance vs. cost framework for evaluating DHT design tradeoffs under churn. In *Proceedings of the 24th INFOCOM*, pages 225–236, 2005. (Cited on page 124.)

[100] Xiang-Yang Li, Yu Wand, and Wen-Zhan Song. Applications of *k*-Local MST for Topology Control and Broadcasting in Wireless Ad Hoc Networks. *IEEE Transactions on Parallel and Distributed Systems*, 15(12):1057–1069, 2004. (Cited on page 206.)

[101] Jial Liang, Rakesh Kumar, and Keith W. Ross. The FastTrack overlay: A measurement study. *Computer Networks*, 50(6):842–858, 2006. (Cited on page 85.)

[102] Wen-Hwa Liao, Yucheng Kao, and Chien-Ming Fan. Data aggregation in wireless sensor networks using ant colony algorithm. *Journal of Network and Computer Applications*, 31(4):387–401, 2008. (Cited on page 159.)

[103] Helena R. Lourenço, Olivier Martin, and Thomas Stützle. Iterated Local Search. In Fred W. Glover and Gary A. Kochenberger, editors, *Handbook of Metaheuristics*, volume 57 of *International Series in Operations Research & Management Science*, chapter 11, pages 321–353. Springer, Heidelberg, 2002. (Cited on pages 22, 40, 93, and 191.)

[104] Eng Keong Lua, Jon Crowcroft, Marcelo Pias, Ravi Sharma, and Steven Lim. A Survey and Comparison of Peer-to-Peer Overlay Network Schemes. *IEEE Communications Surveys & Tutorials*, 7: 72–93, 2005. (Cited on pages 112 and 138.)

[105] Olivier Martin, Steve W. Otto, and Edward W. Felten. Large-step Markov chains for the TSP incorporating local search heuristics. *Operations Research Letters*, 11(4):219–224, 1992. (Cited on page 22.)

[106] MaxMind Inc. GeoLite City. Website, 2007. URL http://www.maxmind.com/app/geolitecity. (Cited on pages 122 and 127.)

[107] Filippo Menczer, Melania Degeratu, and W. Nick Street. Efficient and Scalable Pareto Optimization by Evolutionary Local Selection Algorithms. *Evolutionary Computation*, 8(2):223–247, 2000. (Cited on page 22.)

[108] Gregor Mendel. Versuche über Pflanzen-Hybriden. *Verhandlungen des naturforschenden Vereines zu Brünn*, 4:3–47, 1866. In German. (Cited on page 17.)

[109] Peter Merz. Advanced Fitness Landscape Analysis and the Performance of Memetic Algorithms. *Evolutionary Computation, Special Issue on Memetic Evolutionary Algorithms*, 12(3):303–326, 2004. (Cited on page 57.)

[110] Peter Merz and Katja Gorunova. Efficient Broadcast in P2P Grids. In *Proceedings of the 5th IEEE/ACM International Symposium on Cluster Computing and the Grid (CCGrid 2005)*, pages 237–242, 2005. (Cited on page 142.)

[111] Peter Merz and Matthias Priebe. A New Iterative Method to Improve Network Coordinates-Based Internet Distance Estimation. In *ISPDC 2007 – 6th International Symposium on Parallel and Distributed Computing*, pages 169–176. IEEE Computer Society, Los Alamitos, 2007. (Cited on page 120.)

[112] Peter Merz and Steffen Wolf. Evolutionary Local Search for Designing Peer-to-Peer Overlay Topologies based on Minimum Routing Cost Spanning Trees. In Thomas Philip Runarsson, Hans-Georg Beyer, Edmund Burke, Juan J. Merelo-Guervós, L. Darrell Whitley, and Xin Yao, editors, *PPSN 2006*, volume 4193

of *LNCS*, pages 272–281. Springer, Heidelberg, 2006. (Cited on pages 47 and 60.)

[113] Peter Merz and Steffen Wolf. TreeOpt: Self-Organizing, Evolving P2P Overlay Topologies Based On Spanning Trees. In Torsten Braun, Georg Carle, and Burkhard Stiller, editors, *Kommunikation in Verteilten Systemen (KiVS 2007) – Industriebeiträge, Kurzbeiträge und Workshops*, pages 231–242. VDE Verlag, Berlin, 2007. (Cited on page 82.)

[114] Peter Merz, Florian Kolter, and Matthias Priebe. Free-Riding Prevention in Super-Peer Desktop Grids. In *Proceedings of the 3rd International Multi-Conference on Computing in the Global Information Technology (ICCGI 2008)*, pages 297–302. IEEE Computer Society, Los Alamitos, 2008. (Cited on page 154.)

[115] Peter Merz, Matthias Priebe, and Steffen Wolf. A Simulation Framework for Distributed Super-Peer Topology Construction Using Network Coordinates. In Didier El Baz, Julien Bourgeois, and François Spies, editors, *16th Euromicro Conference on Parallel, Distributed and Network-based Processing*, pages 491–498. IEEE Computer Society, Los Alamitos, 2008. (Cited on page 135.)

[116] Peter Merz, Matthias Priebe, and Steffen Wolf. Super-Peer Selection in Peer-to-Peer Networks using Network Coordinates. In *Proceedings of the 3rd International Conference on Internet and Web Applications and Services (ICIW 2008)*, pages 385–390. IEEE Computer Society, Los Alamitos, 2008. (Cited on page 133.)

[117] Peter Merz, Jan Ubben, and Matthias Priebe. On the Construction of a Super-Peer Topology underneath Middleware for Distributed Computing. In *Proceedings of the 8th IEEE International Symposium on Cluster Computing and the Grid (CCGRID 2008)*, pages 590–595. IEEE Computer Society, Los Alamitos, 2008. (Cited on page 154.)

[118] Peter Merz, Steffen Wolf, Dennis Schwerdel, and Matthias Priebe. A Self-Organizing Super-Peer Overlay with a Chord Core for Desktop Grids. In Karin Anna Hummel and James P. G. Sterbenz, editors, *Proceedings of the 3rd International Workshop on Self-Organizing Systems (IWSOS)*, volume 5343 of *LNCS*, pages 23–34. Springer, Heidelberg, 2008. (Cited on page 152.)

[119] Peter Merz, Florian Kolter, and Matthias Priebe. A Distributed Reputation System for Super-Peer Desktop Grids. *IARIA International Journal on Advances in Security*, 2(1):30–41, 2009. (Cited on page 154.)

[120] Wil Michiels, Emile H. L. Aarts, and Jan Korst. *Theoretical Aspects of Local Search*. Monographs in Theoretical Computer Science. Springer, 2007. ISBN 978-3-540-35853-4. (Cited on page 19.)

[121] Alper T. Mızrak, Yuchung Cheng, Vineet Kumar, and Stefan Savage. Structured Superpeers: Leveraging Heterogeneity to Provide Constant-Time Lookup. In *Proceedings of the IEEE Workshop on Internet Applications 2003*, pages 104–111, Los Alamitos, USA, 2003. IEEE Press. (Cited on page 138.)

[122] Nenad Mladenović and Pierre Hansen. Variable Neighborhood Search. *Computers & Operations Research*, 24(11):1097–1100, 1997. (Cited on pages 20, 21, and 94.)

[123] Roberto Montemanni and Luca Maria Gambardella. Exact algorithms for the minimum power symmetric connectivity problem in wireless networks. *Computers & Operations Research*, 32(11): 2891–2904, 2005. (Cited on pages 190, 191, 192, 199, 204, 205, and 214.)

[124] Roberto Montemanni, Luca Maria Gambardella, and Arindam K. Das. The Minimum Power Broadcast problem in Wireless Networks: a Simulated Annealing approach. *Wireless Communications and Networking Conference (WCNC)*, 4:2057–2062, 2005. (Cited on pages 164, 167, 177, 180, and 228.)

[125] Alberto Montresor. A Robust Protocol for Building Superpeer Overlay Topologies. In *Proceedings of the 4th International Conference on Peer-to-Peer Computing*, pages 202–209, 2004. (Cited on page 111.)

[126] Pablo Moscato. On Evolution, Search, Optimization, Genetic Algorithms and Martial Arts: Towards Memetic Algorithms. Caltech Concurrent Computation Program, C3P Report 826, California Institute of Technology, Pasadena, USA, 1989. (Cited on pages 21 and 40.)

[127] Stephen Naicken, Anirban Basu, Barnaby Livingston, and Seth-alat Rodhetbhai. A Survey of Peer-to-Peer Network Simulators. In *Proceedings of The Seventh Annual Postgraduate Symposium*, Liverpool, UK, 2006. (Cited on pages 123 and 124.)

[128] Stephen Naicken, Barnaby Livingston, Anirban Basu, Sethalat Rodhetbhai, Ian Wakeman, and Dan Chalmers. The State of Peer-to-Peer Simulators and Simulations. *Computer Communication Review*, 37(2):95–98, 2007. (Cited on pages 122 and 123.)

[129] John A. Nelder and Roger Mead. A simplex method for function minimization. *Computer Journal*, 7:308–313, 1965. (Cited on page 120.)

[130] T. S. Eugene Ng and Hui Zhang. Predicting Internet Network Distance with Coordinates-Based Approaches. In *INFOCOM 2002. Twenty-First Annual Joint Conference of the IEEE Computer and Communications Societies. Proceedings. IEEE*, pages 170–179, 2002. (Cited on pages 120, 121, and 133.)

[131] Morton E. O'Kelly. A quadratic integer program for the location of interacting hub facilities. *European Journal of Operational Research*, 32(3):393–404, 1987. (Cited on pages 87, 91, 99, and 221.)

[132] Morton E. O'Kelly, Darko Skorin-Kapov, and Jadranka Skorin-Kapov. Lower Bounds for the Hub Location Problem. *Management Science*, 41(4):713–721, 1995. (Cited on pages 90, 91, 100, and 127.)

[133] Morton E. O'Kelly, Deborah L. Bryan, Darko Skorin-Kapov, and Jadranka Skorin-Kapov. Hub Network Design with Single and Multiple Allocation: A Computational Study. *Location Science*, 4(3):125–138, 1996. (Cited on pages 99 and 221.)

[134] Sigurður Ólafsson and Leyuan Shi. A Method for scheduling in parallel manufacturing systems with flexible resources. *IIE Transactions*, 32(2):135–146, 2000. (Cited on pages 27 and 161.)

[135] Venkata N. Padmanabhan and Lakshminarayanan Subramanian. An Investigation of Geographic Mapping Techniques for Internet Hosts. In *SIGCOMM'01: Proceedings of the 2001 conference on Applications, technologies, architectures, and protocols for computer communications*, pages 173–185, New York, NY, USA, 2001. ACM Press. (Cited on page 121.)

[136] Linda Pagli and Giuseppe Prencipe. Brief Annoucement: Distributed Swap Edges Computation for Minimum Routing Cost Spanning Trees. In *Principles of Distributed Systems*, volume 5923 of *LNCS*, pages 365–371, 2009. (Cited on page 62.)

[137] Christos H. Papadimitriou and Kenneth Steiglitz. *Combinatorial Optimization: Algorithms and Complexity*. Dover Publications Inc., Mineola, USA, 1998. ISBN 0-486-40258-4. (Cited on page 12.)

[138] Joongseok Park and Sartaj Sahni. Power Assignment For Symmetric Communication In Wireless Networks. In *Proceedings of the 11th IEEE Symposium on Computers and Communications (ISCC)*, pages 591–596, Washington, 2006. IEEE Computer Society, Los Alamitos. (Cited on pages 189, 192, and 195.)

[139] David Peleg and Eilon Reshef. Deterministic Polylog Approximation for Minimum Communication Spanning Trees. In *ICALP'98: Proceedings of the 25th International Colloquium on Automata, Languages and Programming*, volume 1443 of *LNCS*, pages 670–681. Springer, Heidelberg, 1998. (Cited on page 38.)

[140] Melquíades Pérez Pérez, Francisco Almeida Rodríguez, and J. Marcos Moreno-Vega. On the Use of Path Relinking for the *p*-Hub Median Problem. In Gottlieb and Raidl, editors, *Proceedings of the 7th European Conference on Evolutionary Computation in Combinatorial Optimization*, volume 3004 of *LNCS*, pages 155–164. Springer, 2004. (Cited on page 92.)

[141] Melquíades Pérez Pérez, Francisco Almeida Rodríguez, and J. Marcos Moreno-Vega. A hybrid VNS-path relinking for the p-hub median problem. In *IMA Journal of Management Mathematics*. Oxford University Press, 2007. (Cited on page 92.)

[142] Charles E. Perkins and Pravin Bhagwat. Highly Dynamic Destination-Sequenced Distance-Vector Routing (DSDV) for Mobile Computers. In *Proceedings of SIGCOMM*, pages 234–244, 1994. (Cited on page 143.)

[143] Peter R. Pietzuch, Jonathan Ledlie, Michael Mitzenmacher, and Margo I. Seltzer. Network-Aware Overlays with Network Coordinates. In *Proceedings of the 1st International Workshop on*

Dynamic Distributed Systems (IWDDS'06), 2006.   (Cited on page 117.)

[144] Matthias Priebe. *On the Design of a Middleware for Super-Peer Desktop Grids.* PhD thesis, University of Kaiserslautern, Germany, Logos Verlag Berlin, 2010. ISBN 978-3-8325-2453-1. (Cited on pages 154 and 219.)

[145] Robert C. Prim. Shortest Connection Networks And Some Generalizations. *Bell System Technical Journal*, 36:1389–1401, 1957. (Cited on pages 6, 13, 37, 162, 188, 189, and 205.)

[146] Günther R. Raidl and Ivana Ljubić. Evolutionary Local Search for the Edge-Biconnectibity Augmentation Problem. Technical Report TR-186-1-02-02, Vienna University of Technology, 2001. (Cited on page 21.)

[147] Sylvia Ratnasamy, Paul Francis, Mark Handley, Richard Karp, and Scott Schenker. A scalable content-addressable network. In *SIGCOMM'01*, pages 161–172. ACM Press, 2001. (Cited on pages 61 and 85.)

[148] Sylvia Ratnasamy, Mark Handley, Richard M. Karp, and Scott Shenker. Topologically-Aware Overlay Construction and Server Selection. In *INFOCOM 2002. Twenty-First Annual Joint Conference of the IEEE Computer and Communications Societies. Proceedings. IEEE*, volume 3, pages 1190–1199, 2002. (Cited on page 112.)

[149] Ingo Rechenberg. *Evolutionsstrategie: Optimierung technischer Systeme nach Prinzipien der biologischen Evolution.* Frommann-Holzboog, Stuttgart, 1973. ISBN 3-7728-0373-3. In German. (Cited on pages 16, 17, and 18.)

[150] Matei Ripeanu, Ian Foster, and Adriana Iamnitchi. Mapping the Gnutella Network. *IEEE Internet Computing Journal*, 6(1):50–57, 2002. (Cited on page 61.)

[151] Antony Rowstron and Peter Druschel. Pastry: Scalable, distributed object location and routing for large-scale peer-to-peer systems. In *IFIP/ACM International Conference on Distributed Systems Platforms (Middleware)*, pages 329–350, 2001. (Cited on pages 33, 61, and 85.)

[152] Paolo Santi. *Topology Control in Wireless Ad Hoc and Sensor Networks*. John Wiley & Sons, Chichester, 2005. ISBN 0470094532. (Cited on pages 7, 9, 156, 158, 159, and 160.)

[153] Mario Schlosser, Michael Sintek, Stefan Decker, and Wolfgang Nejdl. HyperCuP - Hypercubes, Ontologies and Efficient Search on Peer-to-Peer Networks. In *Agents and Peer-to-Peer Computing, First International Workshop, AP2PC 2002*, volume 2530 of *Lecture Notes in Computer Science*, pages 112–124. Springer, Heidelberg, 2002. (Cited on page 85.)

[154] Jawwad Shamsi, Monica Brockmeyer, and Liya Abebe. TACON: Tactical Construction of Overlay Networks. In *Proceedings of the IEEE Global Telecommunication Conference (GLOBECOM 2005)*, pages 926–931, 2005. (Cited on page 112.)

[155] Leyuan Shi and Sigurður Ólafsson. Nested Partitions Method for Global Optimization. *Operations Research*, 48(3):390–407, 2000. (Cited on pages 26, 161, 165, and 168.)

[156] Leyuan Shi and Sigurður Ólafsson. Nested Partitions Method for Stochastic Optimization. *Methodology and Computing in Applied Probability*, 2(3):271–291, 2000. (Cited on page 161.)

[157] Leyuan Shi, Sigurður Ólafsson, and Ning Sun. New parallel randomized algorithms for the traveling salesman problem. *Computers and Operations Research*, 26(4):371–394, 1999. (Cited on page 27.)

[158] Leyuan Shi, Sigurður Ólafsson, and Quon Chen. An optimization framework for product design. *Management Science*, 47(12): 1681–1692, 2001. (Cited on pages 27 and 161.)

[159] Sameh Al-Shihabi. Ants for Sampling in the Nested Partition Algorithm. In Christian Blum, Andrea Roli, and Michael Sampels, editors, *Hybrid Metaheuristics*, pages 11–18, 2004. (Cited on page 27.)

[160] Sameh Al-Shihabi, Peter Merz, and Steffen Wolf. Nested Partitioning for the Minimum Energy Broadcast Problem. In Vittorio Maniezzo, Roberto Battiti, and Jean-Paul Watson, editors, *LION 2007 II. Selected Papers*, volume 5313 of *LNCS*, pages 1–11. Springer, Heidelberg, 2008. (Cited on pages 186 and 187.)

[161] Kazuyuki Shudo, Yoshio Tanaka, and Satoshi Sekiguchi. Overlay Weaver: An overlay construction toolkit. *Computer Communications*, 31(2):402–412, 2008. (Cited on page 124.)

[162] Alok Singh. A New Heuristic for the Minimum Routing Cost Spanning Tree Problem. In *International Conference on Information Technology (ICIT'08)*, pages 9–13, Los Alamitos, USA, 2008. IEEE Computer Society. (Cited on page 39.)

[163] Darko Skorin-Kapov and Jadranka Skorin-Kapov. On tabu search for the location of interacting hub facilities. *European Journal of Operational Research*, 73:502–509, 1994. (Cited on page 91.)

[164] Ahmed Sobeih, Jun Wang, and William Yurcik. Performance Evaluation and Comparison of Tree and Ring Application-Layer Multicast Overlay Networks. In *Proceedings of ICENCO'04*, pages 113–118, Cairo, Egypt, 2004. (Cited on page 61.)

[165] Jinhyeon Sohn and Sungsoo Park. A linear program for the two-hub location problem. *European Journal of Operational Research*, 100:617–622, 1997. (Cited on page 223.)

[166] Jinhyeon Sohn and Sungsoo Park. The single allocation problem in the interacting three-hub network. *Networks*, 35(1):17–25, 2000. (Cited on pages 221 and 223.)

[167] Ion Stoica, Robert Morris, David Karger, M. Frans Kaashoek, and Hari Balakrishnan. Chord: A Scalable Peer-to-peer Lookup Service for Internet Applications. In *SIGCOMM'01*, pages 149–160. ACM Press, 2001. (Cited on pages 33, 61, 85, 135, 138, 142, 143, and 145.)

[168] Jeremy Stribling. PlanetLab All-Pairs-Pings. Website, MIT Computer Science and Artificial Intelligence Laboratory, 2003–2005. URL http://pdos.csail.mit.edu/~strib/pl_app/. (Cited on page 49.)

[169] Thomas Stützle. Iterated Local Search for the Quadratic Assignment Problem. Technical Report AIDA-99-03, Technische Hochschule Darmstadt, 1999. (Cited on page 22.)

[170] Thomas Stützle and Holger H. Hoos. $\mathcal{MAX}-\mathcal{MIN}$ Ant System. *Journal of Future Generation Computer Systems*, 16:889–914, 2000. (Cited on pages 25, 26, and 169.)

[171] Su-Wei Tan, A. G. Waters, and John Crawford. MeshTree: Reliable Low Delay Degree-Bound Multicast Overlays. In *International Conference on Parallel and Distributed Systems*, volume 2, pages 565–569, Los Alamitos, USA, 2005. IEEE Computer Society. (Cited on page 62.)

[172] Andrew S. Tanenbaum. *Computer Networks*. Prentice Hall, 3rd edition, 1996. ISBN 978-0133499452. (Cited on page 28.)

[173] Gerard Tel. *Introduction to Distributed Algorightm*. Cambridge University Press, 1994. ISBN 0-521-47069-2. (Cited on pages 28 and 29.)

[174] Madeleine Theile. Exact Solutions to the Traveling Salesperson Problem by a Population-Based Evolutionary Algorithm. In Carlos Cotta and Peter Cowling, editors, *EvoCOP 2009 – Ninth European Conference on Evolutionary Computation in Combinatorial Optimization*, volume 5482 of *LNCS*, pages 145–155. Springer, Heidelberg, 2009. (Cited on page 17.)

[175] Peng-Jun Wan, Gruia Călinescu, Xiang-Yang Li, and Ophir Frieder. Minimum-Energy Broadcasting in Static Ad Hoc Wireless Networks. *Wireless Networks*, 8(6):607–617, 2002. (Cited on page 162.)

[176] Bin Wang and Sandeep K. S. Gupta. S-REMiT: A Distributed Algorithm for Source-based Energy Efficient Multicasting in Wireless Ad Hoc Networks. In *Global Telecommunications Conference, GLOBECOM 2003*, pages 3519–3524. IEEE, 2003. (Cited on page 206.)

[177] Jeffrey E. Wieselthier, Gam D. Nguyen, and Anthony Ephremides. On the Construction of Energy-Efficient Broadcast and Multicast Trees in Wireless Networks. In *Proceedings of the 19th IEEE INFOCOM 2000*, pages 585–594, 2000. (Cited on pages 162, 168, and 173.)

[178] Jeffrey E. Wieselthier, Gam D. Nguyen, and Anthony Ephremides. Energy-efficient broadcast and multicast trees in wireless networks. *Mobile Networks and Applications*, 7(6):481–492, 2002. (Cited on page 162.)

[179] Jeffrey E. Wieselthier, Gam D. Nguyen, and Anthony Ephremides. Distributed algorithms for energy-efficient broadcasting in ad hoc networks. In *Proceedings of IEEE MILCOM 2002*, volume 2, pages 820–825, Anaheim, USA, 2002. (Cited on page 206.)

[180] Steffen Wolf. On the Complexity of the Uncapacitated Single Allocation *p*-Hub Median Problem with Equal Weights. Internal Report 363/07, University of Kaiserslautern, Kaiserslautern, Germany, 2007. (Cited on page 223.)

[181] Steffen Wolf and Peter Merz. Evolutionary Local Search for the Super-Peer Selection Problem and the *p*-Hub Median Problem. In Thomas Bartz-Beielstein, María José Blesa Aguilera, Christian Blum, Boris Naujoks, Andrea Roli, Günther Rudolph, and Michael Sampels, editors, *HM 2007 – 4th International Workshop on Hybrid Metaheuristics*, volume 4771 of *LNCS*, pages 1–15. Springer, Heidelberg, 2007. (Cited on pages 92 and 107.)

[182] Steffen Wolf and Peter Merz. Evolutionary Local Search for the Minimum Energy Broadcast Problem. In Jano van Hemert and Carlos Cotta, editors, *EvoCOP 2008*, volume 4972 of *LNCS*, pages 61–72. Springer, Heidelberg, 2008. (Cited on page 186.)

[183] Steffen Wolf and Peter Merz. Iterated Local Search for Minimum Power Symmetric Connectivity in Wireless Networks. In Carlos Cotta and Peter Cowling, editors, *EvoCOP 2009 – Ninth European Conference on Evolutionary Computation in Combinatorial Optimization*, volume 5482 of *LNCS*, pages 192–203. Springer, Heidelberg, 2009. (Cited on page 204.)

[184] Steffen Wolf and Peter Merz. Efficient Cycle Search for the Minimum Routing Cost Spanning Tree Problem. In Peter Cowling and Peter Merz, editors, *EvoCOP 2010 – Tenth European Conference on Evolutionary Computation in Combinatorial Optimization*, volume 6022 of *LNCS*, pages 276–287. Springer, Heidelberg, 2010. (Cited on page 60.)

[185] Steffen Wolf, Tom Ansay, and Peter Merz. A Distributed Range Assignment Protocol. In Thrasyvoulos Spyropoulos and Karin Anna Hummel, editors, *IWSOS 2009*, volume 5918 of *LNCS*, pages 226–231. IFIP International Federation for Information Processing, 2009. (Cited on page 213.)

[186] Bernhard Wong, Aleksandrs Slivkins, and Ewin Gün Sirer. Meridian: A Lightweight Network Location Service without Virtual Coordinates. In *Proceedings of the 2005 Conference on Applications, Technologies, Architectures, and Protocols for Computer Communications*, pages 85–96. ACM Press, 2005. (Cited on pages 71 and 125.)

[187] Richard T. Wong. Worst-Case Analysis of Network Design Problem Heuristics. *SIAM Journal on Algebraic and Discrete Methods*, 1(1):51–63, 1980. (Cited on page 38.)

[188] Bang Ye Wu and Kun-Mao Chao. *Spanning Trees and Optimization Problems*. Discrete Mathematics and its Applications. Chapman & Hall/CRC, 2004. ISBN 1-58488-436-3. (Cited on page 35.)

[189] Bang Ye Wu, Giuseppe Lancia, Vineet Bafna, Kun-Mao Chao, R. Ravi, and Chuan Yi Tang. A Polynomial-Time Approximation Scheme for Minimum Routing Cost Spanning Trees. *SIAM Journal of Computing*, 29(3):761–778, 1999. (Cited on page 36.)

[190] Bang Ye Wu, Kun-Mao Chao, and Chuan Yi Tang. Approximation Algorithms for Some Optimum Communication Spanning Tree Problems. *Discrete Applied Mathematics*, 102(3):245–266, 2000. (Cited on pages 36 and 65.)

[191] Beverly Yang and Hector Garcia-Molina. Designing a Super-Peer Network. In *Proceedings of the 19th International Conference on Data Engineering*, pages 49–62, 2003. (Cited on pages 85 and 138.)

[192] Ben Y. Zhao, Ling Huang, Jeremy Stribling, Sean C. Rhea, Anthony D. Joseph, and John Kubiatowicz. Tapestry: A Resilient Global-scale Overlay for Service Deployment. *IEEE Journal on Selected Areas in Communications*, 22:41–53, 2004. (Cited on page 85.)

[193] Han Zheng, Eng Keong Lua, Marcelo Pias, and Timothy G. Griffin. Internet Routing Policies and Round-Trip-Times. In *Proceedings of the Passive and Active Network Measurement Workshop*, pages 236–250, 2005. (Cited on pages 118 and 126.)

[194] Stefan Zöls, Zoran Despotovic, and Wolfgang Kellerer. On hierarchical DHT systems – An analytical approach for optimal designs. *Computer Communications*, 31(3):576–590, 2008. (Cited on page 138.)